TIME:
AN ESSAY

For my friends
Johan and Maria Goudsblom

Norbert Elias

Time: An Essay

Translated in part from the German by
Edmund Jephcott

BLACKWELL
Oxford UK & Cambridge USA

This translation copyright © Basil Blackwell 1992

First published in German as *Über die Zeit* by Suhrkamp Verlag, copy © The Norbert Elias Stichting 1987.

Blackwell Publishers
108 Cowley Road
Oxford OX4 1JF
UK

Three Cambridge Center
Cambridge, Massachusetts 02142
USA

All rights reserved. Except for the quotation of short passages for the purposes of criticism and review, no part may be reproduced, stored in a retrieval system, or transmitted, in any form or by any means, electronic, mechanical, photocopying, recording or otherwise, without the prior permission of the publisher.

Except in the United States of America, this book is sold subject to the condition that it shall not, by way of trade or otherwise, be lent, resold, hired out, or otherwise circulated without the publisher's prior consent in any form of binding or cover other than that in which it is published and without a similar condition including this condition being imposed on the subsequent purchaser.

British Library Cataloguing in Publication Data

A CIP catalogue record for this book is available from the British Library.

Library of Congress Cataloging-in-Publication Data

Elias, Norbert.
    Time : an essay / by Norbert Elias.
       p. cm
    Includes index.
    ISBN 0-631-15798-0 (alk. paper)
    1. Time. I. Title.
BD638.E42 1992   91-32598
115-dc20   CIP

Typeset in 12 on 14 pt Sabon
by Tecset Ltd, Wallington, Surrey
Printed in Great Britain by Billing & Sons Ltd, Worcester

This book is printed on acid-free paper

Contents

Preface   1

Time: An Essay   37

Notes   201

Index   209

Preface

'I know what time is if I am not asked', a wise old man once said, 'if I am asked, I no longer know.' Why do I ask?

By exploring problems of time one can find out a good deal about human beings and about oneself that was not properly understood before. Problems in sociology and the human sciences in general that were not clarified by previous theories now become accessible.

Physicists sometimes say that they measure time. They use mathematical formulae in which the measure of time appears as a specific quantum. But time can be neither seen nor felt, neither heard nor tasted nor smelt. This is a question that still awaits an answer. How can something be measured that is not perceptible to the senses?

But do not clocks measure time? They can certainly be used to measure something. This something however, is not really invisible, intangible time but something very tangible, such as the length of a working day or of a lunar eclipse, or a sprinter's speed in the 100 metres.

Clocks are themselves sequences of physical events. They can serve as social norms. They can be divided into evenly recurring patterns such as hours or minutes. If the state of social development requires and permits, these patterns can be identical across a whole country or even many

I am indebted to Artur Bogner for his help in preparing this preface.

countries. So with the aid of clocks the speed of aircraft flying the same distance in quite different places can be compared. With their help one can compare the duration and speed of perceptible processes which by their nature, as they happen successively, allow no direct comparison, such as the length of two speeches when one follows the other. When people find it necessary to do so they use a socially standardized sequence in order to compare sequences that are not directly comparable. Why people do so, and at which stage of development they evolve a unifying concept at so high a level of synthesis as that of 'time', is a question that requires further research. Whatever the answer, time represents common features of observable sequences which people wish to grasp by referring them to a standard sequence. The following study is at least a step in exploring a problem which has so far received less attention than it deserves. How do people advance from a means of orientation at a relatively low level of synthesis to a means of orientation at a relatively high level?

What one can be sure of is the fact that clocks, like simple natural processes with the same social function, serve human beings as means of orienting themselves within the succession of social, biological and physical processes in which they find themselves placed. They also serve people as a way of regulating their behaviour in relation to each other and themselves. They help to coordinate it with other behaviour and with natural processes, that is, processes not influenced by human beings.

If at earlier stages of development the need arose for people to find an answer to questions regarding the location or duration of happenings in the general succession of events, they usually took a specific type of natural process as the standard sequence. They confined themselves to natural processes such as the changing of seasons or the migration of birds. In reality each of these events was unique like our calendar years, but they recurred with sufficient regularity to serve as a ready-made yardstick for recurrent social activities. Such a recurrent pattern, for example, the ebb and flow of the tide, the beat of their own pulse or the coming and going of the sun and moon, could be used by people at an earlier stage as a means of co-ordinating their activities with each other and with extra-human events in the same way that people at a later stage use the recurring symbolic patterns on human-made clockfaces. In the sociology of knowledge up to now the development of knowledge, that is, of human means of orientation, has been somewhat neglected. Moreover, the question how people learn to orient themselves in their world, more and more successfully in the course of millennia, is certainly of no small importance to them in understanding themselves. The development of the measurement of time as a means of orientation in the unceasing flow of events is one example of this. The high social significance of the physical sciences in our age has contributed to a situation in which time is regarded somewhat self-evidently as a datum belonging to the great complex of non-human natural events and so as an object of scientific investigation within the framework of physics.

A glance at the development of the means of measuring time, or timing, shows that this pre-eminence of physics and the naturalist conception of time are relatively recent phenomena. Until Galileo's time what we call 'time', and even what we call 'nature' were centred primarily on human groups. Time was above all a means of orientation in the social world, of regulating the communal life of human beings. The natural sequences elaborated and standardized by people were used as means of determining the location or duration of social activities in the flow of events. Only in more recent times did the use of clocks as important instruments for investigating purely natural processes branch off from this social stream. As this happened the mystery of time, that had long been felt, was deepened. Here we touch on a fundamental problem of sociology: something arises from the communal life of humankind which they do not understand, which seems enigmatic and mysterious. That clocks are instruments constructed and used by people for quite specific purposes connected with the demands of their communal life is understood readily enough. But that time too has an instrumental character is obviously more difficult to grasp. Do we not feel it passing ineluctably over our heads? Linguistic usage also clouds the issue; it makes it appear as if time were a mysterious something the measure of which is determined by man-made instruments, clocks.

How far the inability to take account of the social functions of time for orientation and regulation has contributed to the difficulty people have encountered in elaborating a generally acceptable theory of time, is seen particularly clearly in the attempts of traditional philosophy to solve this problem. At the centre of the long philosophical debate on the nature of time there were – and perhaps still are – two diametrically opposed positions. On one side was the notion that time exists objectively as a part of natural creation. Its mode of being, according to this view, was not different to that of other objects of nature, except that it was not perceptible. Newton was perhaps the most prominent representative of this objectivist conception, which in more recent times has found itself at a disadvantage. In the opposing camp the notion prevailed that time was a kind of synoptic view of events residing in the peculiar nature of human consciousness, or the human mind or reason, which therefore preceded all human experience as its condition. Descartes already inclined towards this view. It found its most influential expression in the philosophy of Kant, who saw time and space as representing an *a priori* synthesis. In less systematic form this view appears now to have largely gained the upper hand. In simple terms it states that time is no more than an innate form of experience, and so an unalterable fact of human nature.

The two opposed theories of time have, as we can see, a number of basic assumptions in common. In both, time is represented as a fact of nature, in one case an 'objective' fact existing independently of human beings and in the other as a merely 'subjective' notion rooted in human nature. In this confrontation between a subjectivist and an objectivist theory of time a fundamental feature of the traditional philosophical thory of knowledge is reflected. It is taken for granted that there is a universal, endlessly repeated starting point of knowlege. A person, so it appears, steps before the world quite alone, a subject before objects, and starts to have knowledge. The only question is whether, in the formation of human conceptions such as the idea that all events are immersed in the flow of time, the nature of the subject or of the object has precedence.

The artificiality of these common assumptions of the opposed traditional theories, and the sterility of the endless debates betweeen their exponents, only emerge in full clarity if one turns unequivocally away from the paradigmatic assumptions of the old theories – not only in relation to time – and opposes to them a theory of human knowledge that stays in close touch with the observable development of human knowledge, so that its paradigmatic assumptions are capable of being tested and revised. The text which follows is a contribution to such an endeavour. Human knowledge is the result of humanity's long learning process, which has no beginning – such is the conception underlying this study. Each human being, no matter how much innovation he may contribute, builds on a pre-existing stock of knowledge which he extends. The case is no different with the knowledge of time.

In the more developed societies it seems almost self-evident, for example, that a person knows how old he is. Members of such societies react with amazement, or perhaps head-shaking incomprehension, when they learn that in simpler societies there are people who are unable to give any definite answer when asked their age. But if a group's social fund of knowledge does not include a calendar it is difficult to determine the number of one's own years. One is unable to compare the length of one period of one's life directly with another. To do so one would need as a framework of reference another sequence with recurrent divisions the length of which is socially standardized. In short, one would need what we call a calendar. The calendar's unrepeatable succession of numbered years symbolically represents the unrepeatable succession of social and natural events. It thus serves as a means of orientation in the great continuum of change which is at once the natural and the social world. The months and numbered days of the calendar represent repeatable patterns of the unrepeatable succession of events. The totality of these calendar symbols is in more complex societies quite indispensable for the regulation of social existence, as in deciding the dates of holidays, the duration of contracts, and so on.

So, in these societies, knowledge of calendar time, as of clock time, as means of orientation in relation to oneself and one's age, is taken for granted to the point where it escapes reflection. One does not ask how it is to be explained that communal life at earlier stages of development is able to function without calendar or clocks, while at later stages it is scarcely possible without these means. One does not ask why and how the differentiation of these standardized implements, often exact to the day, the hour or even the second, has come into being, or the corresponding pattern of individual self-control, the self-imposed compulsion to know the time. No urgent need is felt as yet by members of a society to discover the connections between a social structure with an indispensable but also inescapable network of temporal definitions, and a personality structure with a very acute and disciplined sensitivity to time. They feel at their backs the pressure of everyday clock time and – more intensely as they grow older – the flight of calendar years. All this becomes second nature; it appears, and is accepted, as the fate of all human beings.

And this blind process goes further in the same direction. Particularly, but not solely, in the case of high positions of co-ordination, the number of chains of interdependence which intersect at these positions is constantly increasing, and with it the pressure on the people occupying them to timetable their professional activity ever more exactly.

If, keeping this highly developed society in mind, we look back at simpler societies, we may better understand how, there too, the macrocosm of the group and the microcosm of the individual are interdependent in their structure and – more or less – co-ordinated. In the case of relatively self-sufficient village states, which may well make war on each other from time to time (examples will be found in the text), the chains of interdependence intersecting in the individual are usually short, few in number and lacking in complexity. The unchanging recurrence of the same sequence patterns, such as the cycle of the seasons, usually looms much larger in the knowledge of people at this stage than the succession of years which never return. And at this stage the individual is usually less sharply distinguished in his own consciousness as a unique person from the chain of generations than happens in highly differentiated societies at a later stage. It may be that the man who says: 'When I built this house . . . ' is speaking of his grandfather. This also explains how people in societies without calendars and so without precise symbols for the unrepeatable succession of years, can lack exact knowledge of their own age. They may say: 'I was a child when the great earthquake happened.' The event which here serves as the framework of reference for when-questions has not yet acquired the character of a continuous process. It exists as discrete points, and the correlation between the event to be dated and the dating framework is a low-level synthesis. Concepts at the level of synthesis of the concept 'time' are still outside the horizon of knowledge and experience.

All this perhaps makes it clearer that what the concept of time refers to is neither a conceptual 'copy' of a flow existing objectively, nor a category of experience common to all people and existing in advance of all experience. Among the difficulties one encounters in reflecting on time is the fact that time does not fit neatly into any of the mental pigeonholes that are still used casually today as means for classifying phenomena of this kind. The problem of time often appears as a problem for physicists and metaphysicians. This has cut the ground from under the feet of anyone who reflects on time. To stand on solid ground it is not enough to contrast time as an object of sociology to time as an object of physics or, in other words, to oppose 'social time', as happens on occasion, to 'physical time'. Dating, and measuring time in general, cannot be understood on the basis of a conception of the world as split into 'subject' and 'object'. Its preconditions are both physical processes, whether untouched or shaped by human beings, and people capable of mental synthesis, of seeing together what does not happen together but successively. Not 'people' and 'nature' as two separate entities but 'people in nature', is the basic concept which is needed in order to understand 'time'. So the endeavour to discover the nature of time helps us to understand that the splitting of the world into 'nature', the sphere of the natural sciences, and human societies, the sphere of the social and human sciences, presents an illusory picture of a split world which is an artificial product of an erroneous development within science.

The meaning of the concept of 'nature' is today determined to a high degree by its relation to the present social form and meaning of the natural sciences. These sciences are concerned with only a limited area of reality; they confine themselves to particular levels of integration within the natural universe, excluding from their sphere the highest levels of integration, represented by human beings, as if they were not part of 'nature'. However, it is of no small significance for what is to be understood as 'nature' that human beings have emerged from the natural universe as representatives of a high, perhaps the highest, stage of differentiation and integration. This fact cannot be ignored; the concept of nature must take account of the circumstance that its blind processes have given rise not only to helium reactors and lunar deserts, but to human beings also. In this connection we do not need to discuss what responsibility this knowledge places on human beings or, more generally, what its implications are in terms of the behaviour of people towards each other and towards 'nature' in the narrower sense of the word. The traditional division of academic subjects implies a notion of a world split into 'nature' and 'society', or into 'nature' and 'culture'. Movements such as environmentalism may announce a growing awareness that people do not exist in isolation but are embedded in the natural process, and that through their special nature, and for their own sake, they bear responsibility for this relationship.

The problems which people try to solve by measuring the 'time' of something, point towards the same basic position, towards groups of people who find themselves placed in a larger world, the natural universe. Wherever one operates with 'time', people within their 'environment', within social and physical processes, are always involved. This book poses the general question of why people need to measure time. The answer is not simple. It often takes us far afield. But one can begin to answer the question by saying: because positions and sequences which have their places successively in the unending flow of events cannot be juxtaposed and directly compared. If for some reason it becomes important for the people of a society to define positions and periods which succeed each other in the flow of events, a second succession of events is needed in which, although all the individual changes succeed each other unrepeatably, certain patterns of change recur with some regularity. Recurrent patterns in this second sequence then serve as standardized reference points by means of which, since they represent the return not of the same thing but of the same interval, events in the other sequence, that cannot be compared directly, can now be compared indirectly. The apparent movement of the sun as it passes through one particular point on the horizon to another, the movements of the hands from one point on a clockface to another, are examples of recurrent patterns that can be used as reference units and means for comparing segments of events in other sequences, the successive segments or positions of which cannot be related together directly. As regulative and cognitive symbols these units thus take on the meaning of time units.

The expression 'time' therefore refers to this relating together of positions or segments within two or more continuously moving sequences of events. The sequences themselves are perceptible. The relation between them results from the elaboration of perceptions by human beings possessing knowledge. It finds expression in a communicable social symbol, the concept of 'time', which within a certain society can transmit from one person to another a memory picture which can be experienced, but not perceived through the senses.

The ingrained habits of the old theory of knowledge might here prompt the question: is time therefore only a relation established by a person, and not something actually existing independently of him or her? This is a false conclusion, resulting in part from positing the subject of knowledge as an individual person. The individual does not invent the concept of time on his own. One learns both the concept and the social institution of time inseparable from it from childhood on, if one grows up in a society in which this concept and this institution are at home. In a society where this is the case one learns the concept of time not only as a tool of reflection, the result of which is intended for publication in philosophical books; every growing child in such societies learns to recognize 'time' as a symbol of a social institution the external compulsion of which the child soon feels at first hand. If he does not learn during the first ten years of life how to evolve an apparatus of self-constraint compatible with this institution, if, in other words, a child in such a society does not learn early on how to regulate his behaviour and feelings in keeping with the social institution of time, it will be very difficult, if not impossible, for such a person to take up the position of an adult in this society.

The conversion of the external compulsion coming from the social institution of time into a pattern of self-constraint embracing the whole life of an individual, is a graphic example of how a civilizing process contributes to forming the social habitus which is an integral part of each individual personality structure. At the same time, to recall this civilisatory conversion of external compulsion by the social institution of time into a kind of temporal conscience (a conversion which is not necessarily trouble-free, as is seen in compulsive unpunctuality, for example) can help in explaining the idea that it is an innate peculiarity of human consciousness that forces each human being to experience everything as taking place within the flow of time. A peculiarity of the social habitus presents itself to reflection as a feature of one's own nature and so of human nature in general. Philosophical subjectivism draws much of its conviction from this misinterpretation of the felt inescapability of one's own experience of time.

If we follow this trail further, it is not very difficult to avoid the traps set by traditional epistemology. A comparison may help; like all comparisons it is lame. The measurement of time has specifiable functions for human beings. In the course of social development these functions can change in ways that can again be specified. The measurement of time and the implements serving it change accordingly. What still remains by and large unclear today is the ontological status of time. People reflect on it without really knowing what kind of phenomenon they are dealing with. Is time a natural phenomenon? Is it a cultural object? Or does the substantive form of the word 'time' perhaps create only an illusion of an object? What do clocks really show when we say that they show the time?

We may perhaps take a step forward if we compare time with an implement, something which people have created 'in the course of time', as the expression goes, and which fulfils quite specific functions. Let us take a boat as an example. It would be surprising if anyone postulated that a boat had the same ontological status as a sea or river, that is, was a natural object. It would be hardly less surprising if someone argued that the boat builder had followed some transcendental concept of a boat in building it, a gift of nature innate in him, independently of any experience of rivers or seas, on which it is the purpose of boats to sail as means of transportation. In the case of a boat it is quite obvious that it is built by people for quite specific purposes. If there were no people there would be no boats. In this respect the comparison holds good. In a world without people or living beings of any kind there would be no time. There would be no clocks and no calendars. But if one says this with regard to time one often passes unawares into a traditional mental world in which time is understood simply as a concept or idea and where the central question addressed to an idea is whether it is a faithful *replica* of reality or not. But is time, considered as an object or reflection, in fact nothing other than a conception of individual people? An answer is to be found in what follows. But thoughts will not stand still. A few further considerations will perhaps be useful.

Like boats and clocks, time is something that has evolved among human beings in relation to quite specific tasks which it fulfils. In the flow of events, as has been said, positions and segments can be defined with the help of 'time' that would otherwise remain indefinable. In the present day one needs 'time' to find one's way successfully in a plethora of diverse tasks. But if one points out that 'time' is a means of orientation created by people, one can easily slip into saying that it is *merely* a human invention. In keeping with the expectation aroused by the replica theory, this 'merely' expresses disappointment at finding time not to be an 'idea' that faithfully mirrors something really existing. But time is not simply an 'idea' that appears from nowhere in the heads of individuals. It is also a social institution varying with the state of social development. In growing up, the individual learns to understand the time signals customary in his society and to regulate his behaviour by them. The memory picture of time, the idea of it that an individual possesses, therefore depends on the state of development of the institutions representing and communicating time, and on the individual's experience of them from an early age.

In more developed societies clocks are among the most prominent of the institutions representing time. But clocks are not time. Time too has an instrumental character, but of a particular kind. Here we come to the peculiar problem of the relation between clock and time. What is the relation of the physical process of a timing device, e.g. the clock mechanism, to the social function of the implement as communicator of time? In its capacity as time measurer the device is a transmitter of messages to human recipients just as newsprint is the physical bearer of messages to newspaper readers. Whether we are concerned with clocks or the apparent circling of the sun around the earth, means of time measurement are always perceptible sequences of events or, as in the case of a calendar, sequences simulated in written or printed form. But they are also something other than physical sequences of events. A physical sequence only takes on the character of a timing device through also possessing, in connection with its physical aspects, the character of a moveable social symbol and participating as such, whether through informing or regulating, in the communications system of human societies. Whatever else they may be, timing devices are always transmitters of messages to people. Clocks are certainly physical sequences constructed by human beings. But they are so constructed that in one way or another, for example, by the changing constellation of the hands on the clockface, they are incorporated in the symbolic world of human beings.

The changing constellation on the face of a clock has the function of showing people the position they and others at present occupy in the great successive flow of events, or how long they have taken to get from one place to another. The human-made symbols of the changing faces of clocks, the changing dates of calendars, *are* time. It is now ten past one. That is time. This requires a certain circumspection in the use of language. One can say rightly enough that the clock communicates time. But it does so by the continuous production of symbols which only have meaning in a five-dimensional world, only in a world in which there are people, and therefore only for beings who have learned how to connect specific memory images, a particular meaning, with the perceptible form. The uniqueness of time lies in the fact that symbols – at the present stage predominantly numerical symbols – are used for orientation in the incessant flow of events; events taking place on all levels of integration, physical, biological, social and individual.

This does not mean that the symbolic aspect of a timing implement such as a clock is separable in any sense from its physical aspects. As, in the case of language, physical sound patterns and memory images fuse inseparably into symbols, in the form of the clock a fifth-dimension characteristic of communication among people is added to the four-dimensional event of a movement in space and time. By the use of a clock a group of people, in a sense, transmits a message to each of its individual members. The physical device is so arranged that it can function as a transmitter of messages and thereby as a means of regulating behaviour within a group. What the clock communicates by the symbols on the clockface is what we call time. One looks at the clock and finds out that it is now such and such a time, not only for me but for the whole of the society to which I belong. The symbols on the clockface inform me in addition about aspects of natural events, for example, the present constellation of sun and earth in the endless succession of their movements. It informs me whether it is now morning or evening, day or night. At the present stage of development time, as we can see, has become a symbol of a very wide network of relationships, in which sequences on the individual, social and non-human natural planes are interconnected.

All this somewhat changes our picture of the relation between individual, society and nature. At present we are still working largely with concepts which draw sharp divisions between the physical, social and individual planes of integration. Particularly as regards the position of the individual in the world, the prevailing picture has gone somewhat out of balance. The individual often appears to stand opposed to the world as an isolated being, and behaves accordingly. Society and nature, too, often appear as separate worlds. Reflection on time can perhaps help to correct this picture of a world with hermetically sealed compartments. Such reflection can make no progress if one refuses to acknowledge that nature, society and individuals are embedded in each other and are interdependent.

A glance at a clock or calendar shows this very vividly. If one finds that it is now twelve o'clock on the twelfth day of the twelfth month of the year 1212, this can serve to define a temporal point in the flow at the same time of an individual life, of society and of natural events. At its present stage of development time, as we can see, is a symbolic synthesis at a very high level, a synthesis by means of which positions in the succession of physical natural events, of the social process and of an individual lifespan can be related together.

Calendar time is a good example of this synthesis. In the symbolic flow of the never-returning years are incorporated the symbolic units of months, weeks and days that recur at particular intervals. The relation between these time symbols and symbols of apparently recurring natural sequences, with which they were originally associated, has not been entirely lost, but is not nearly as close as it was at earlier stages of development. Ways have been found to smooth out, to an extent, the irregularities in the relation between the movements of heavenly bodies like the sun and the moon. These were too great for the growing social need for maximum regularity in the temporal sequence. But the task of harmonizing natural sequences and the demands arising from social sequences in the form of calendar time was anything but simple. It took several thousand years for people to learn how to produce calendars in which the human representation of time in the symbolic form of recurring time units needed for the regulation of social events, and the natural sequences on which the symbolic representation was modelled, did not sooner or later fall out of step. Perhaps the self-evident character of time in developed societies has to do with the fact that timing devices like calendars now function relatively smoothly and unproblematically.

Within a person's society, within the network comprised by its members, the individual usually has a certain measure of autonomy, some scope for personal decision. Human beings have a degree of autonomy, of scope for decision, within uncontrolled, non-human nature. But this scope, that can grow or contract, has limits. Undoubtedly, uncontrolled natural events always have the last word, but from them have now emerged living beings at the highest level of order yet known, human beings who possess among other things the unique capacity to communicate through the use of symbols that are not genetically fixed but created by humans, that are learned and society-specific, by means of which people can find their bearings in the world.

In referring to the symbolic character of time, therefore, it may be useful to mention a circumstance that can only be touched on in passing. The dominant form of human communication is through social symbols. The language of a group becomes the individual's learned means of communication in growing up; it becomes his language, an integral part of the individual person. Among the peculiarities of the human society of the 'many', in other words, is the circumstance that its manifestations are not only something existing outside the individual person, not only in the 'external world', but are at the same time structural features of the 'individual'. The ever-repeated transformation of the social language into an individual language is only one of many examples of this constant individualization of social facts. It is frequently overlooked or obscured by its perspective correlative, the socialization of the individual.

In this latter concept is reflected an image of humanity which places the 'individual' at the centre and presents the 'many' as something added later. Communal life with others and its demands are envisaged under the concept of socialization as something supplementary to the individual. In this case the approach from the standpoint of the individual appears in a relatively mild and partly acceptable form. There are other cases when this is not so. The theories of action going back to Max Weber are an example. As in the case of geocentric theories, it is understandable that, without further reflection, one can take oneself, just as one finds oneself, and in generalized form the 'individual human being', as the starting point in thinking about the world. The individual person appears from this viewpoint as the beginning, from which the communal life of people in societies can be explained. This kind of sociological theorizing starting from the individual takes on a special timbre when the individual person is reduced in thought to the individual action. Society then appears really as a mosaic of individual actions by individual people, calling to mind the lines of Morgenstern:

> A knee walks lonely through the world
> It is a knee and nothing else!

So, it appears, do actions walk alone in the world. Or, who knows, perhaps the individual actions form swarms together and so societies.

It would scarcely enhance our understanding even of a whole person to suppose that he first walked alone in the world and then, as if retrospectively and by chance, adjusted his behaviour to that of other people. Each human being implies others who were there before him. A child develops into a human being only by becoming part of a group, for example by learning a language which was there before him, or by acquiring a civilizatory canon of instinct and affect control. This is not only indispensable for communal life with others, but also for living with oneself, for developing into a human individual and for survival.

The plurality of human beings represents an order of a special kind. The uniqueness of human communal life brings into being unique, specifically social facts which cannot be understood or explained by thinking from the individual standpoint. Language is a good example. How would it be if one awoke one morning to find all other people speaking a language that one did not understand? Language illustrates in paradigmatic form a type of social fact which is presupposed by a plurality of people and which is individualized over and over again. Such facts are, as it were, precipitated in each new member of a group and become the patterns of his personal behaviour and feelings, his social habitus, from which develops that which distinguishes the individual member of the group from others. The common language pattern can, to an extent, be adapted individually. But if individualization goes too far, the language loses its function as a means of communication for a group.

The formation of conscience, the patterns of instinct and affect control, are other examples. Others are money and time. Corresponding to all of them are personal patterns of feeling and behaviour, a social habitus which the individual shares with others and which forms an integral part of the individual personality structure. For example, in societies which exert strong pressure for a scrupulous use of money, a whole welter of individual variations is to be found, testifying to difficulties in the individualization of social patterns. One finds people who at every possible opportunity, following a kind of personal compulsion, borrow small sums and forget to return them, others who compulsively live 'beyond their means'. Similarly, in a society with a high standard of individual self-control as regards time, one can observe over and again individuals who are compulsively unpunctual.

It is perhaps still somewhat difficult at present to visualize the emergence of an individual habitus from a social one. This unique characteristic of human societies, the fact that it is only possible for an individual to develop into a relatively autonomous person with a personality structure which is individually stamped and in some respects more or less unique, by learning from others, by assimilating social patterns of self-control – this seemingly paradoxical situation is not easy to formulate with the concepts at present available. It can probably be understood if one says that language is a means of socialization. But this would not fully do justice to the peculiarity of this learning process. In humans the natural provision of unlearned means of communication is far less than in other organisms. This results in a far greater dependence of individuals on the biologically prepared acquisition of a social means of communication, a language, through learning. As the means of communication of many, a language is always already in existence when the individual learns it. It is therefore not quite inappropriate to say in this case that the language individualizes itself.

Moreover, the uniqueness of the language symbols which evolve in use by human groups is not confined to their function as means of human communication. I can only point out in passing that among human beings society-specific symbols have also acquired the function of an indispensable means of orientation, that is, the function of knowledge. Like language – and in the form of linguistic symbols – a stock of knowledge transmissible from generation to generation is always present in a human group before the individual enters it, and his preparatory growth potential then makes possible, in the form of learning, the individualization of social knowledge. The fact that people must and can orientate themselves in their world by acquiring knowledge, their total dependence on the learning of social symbols for their survival as individuals and as a group, is another of the peculiarities distinguishing humans from other living beings.

Among the symbols which human beings can learn and, from a certain stage of social development on, must learn as means of orientation, is time. Here too one can speak of the individualization of a social fact. But the self-regulation of the individual speaker in conformity with the group language is a feature of the whole human species, the result, we must suppose, of a long evolutionary process in hominids. The self-control of people in conformity with time comes into being very gradually in the course of human development. Only at relatively late stages of this development does 'time' become the symbol of an inescapable and all-embracing compulsion. In the course of this study it may become somewhat clearer that this compulsion of time comprises both a pressure exerted by the 'many' on individuals, a social compulsion, and at the same time natural pressures, such as ageing. Thinking from the 'many' to the individual, thinking in terms of 'figurations', is perhaps somewhat difficult at present. It requires a level of self-detachment which by and large has not yet been attained. In particular, an ideological commitment can impede understanding. Reference to the necessity, in investigations on the human level, to proceed in thought from the 'many' to the individual, might be taken as expressing a political conviction, a higher valuation of 'society' as against the 'individual'. But we are not concerned here with the question what the relation between the 'individual' and 'society' *ought* to be. We are concerned with what the relation *is*. In other words, we are dealing with a diagnostic problem.

Seen in this way as a problem requiring the investigation of facts and therefore as a means of sociological diagnosis, the emergence of the time-compulsion, on which more is to be read in the text of this book, and the relation between 'society' and 'individual' encountered in this context, are particularly illuminating. That time is experienced as a mysterious power exerting constraint on oneself is certainly nothing new. In Horace we read:

> Eheu fugaces, Postume, Postume,
> labuntur anni...[1]
> [Alas! Posthumus, Posthumus,
> The fleeting years glide away...]

The poet's lament at the flow of passing years comes down to us in his verses with undiminished freshness from antiquity. He already attributed to the human-made time symbols, the years, the properties of flowing and transience which are really attributes of the natural event, socially ordered through the regulative symbols, of the individual life in its passage towards death. As long as there have been human beings, and no doubt also for their non-human ancestors, life followed the ever-recurring course from life to death. The compulsive nature of this course and the sequence of its phases did not depend on the will or consciousness of the human beings. But the ordering of the sequence in the form of years was only possible once people had developed for their own purposes the regulative symbol of the year.

The social need for time measurement in ancient societies was far less acute and pervasive than in the more highly organised states of modern times, not to mention the present-day industrial state. In conjunction with a shift towards increased differentiation and integration, in many modern societies a particularly complex system of self-regulation has developed within individual people as regards time, with a correspondingly acute individual sensibility towards time. The external, social compulsion of time, represented by clocks, calendars or timetables, possesses to a high degree, in these societies, the characteristics which promote the formation of individual self-constraints. The pressure of these external constraints is relatively unobtrusive, moderate, even and without violence, but it is at the same time omnipresent and inescapable.

The early stage at which the social regulation of time begins to be individualized in this sense no doubt contributes much to the solidity and inescapability of the personal time-conscience. The inner voice asking the time is ever-present. It is no wonder that, to people with this personality structure, the experience of all natural, social and personal sequences in terms of the regulative time symbols of their societies often appears as a feature of their own nature, and then of human nature in general. People equipped with such an ingrained, uniform and omnipresent time-conscience find it difficult to imagine that there are others who lack the ever-alert compulsion to know the time. This individualization of social time-control therefore bears in almost paradigmatic form the features of a civilizing process.

There is another factor to consider. When, in the course of their development, symbols have attained a very high degree of adequacy to reality, it is often difficult at first for people to distinguish between symbol and reality. In our day, time symbols such as the calendar, while no doubt capable of improvement, have by and large reached a very high level of adequacy, higher than ever before. This too, for many people, easily blurs the distinction between a sequence of events such as their lives on one hand and the relation established by human beings between this sequence and that of the calendar on the other. In such a situation many people cannot resist feeling that time is passing, whereas in reality it is the natural sequence of their life, or perhaps the transformation of society and nature, to which the feeling of passing applies.

Undoubtedly, the capacity of people to orient themselves in their world and to harmonize their behaviour with the aid of regulative symbols is one level of reality. In this sense it is misleading to speak of a greater or lesser congruence of symbols to reality. But here it is perhaps enough to point out that communication by people through society-specific symbols is made necessary and possible by their nature. The question, 'What conclusions concerning the symbol-theory of knowledge and perception follow from the realization that in cognition too, in the meeting of the symbolic with the non-symbolic worlds, metaphorically speaking, nature encounters itself' can be left open. The notion of the old theory of knowledge, that the 'subject' enters a world of 'objects' similar to an alien realm by a kind of ontological accident, now lies behind us.

What deserves further consideration is the diagnostic significance of the relatively high level of individual self-regulation in terms of social time characteristic of people in developed industrial states as a hallmark of the direction of a civilizing process. Perhaps it only becomes quite clear that this ever-wakeful and universal sensitivity to time is a symptom of a civilizing process if this social habitus is compared to that of people in simpler societies with lower social demands concerning time. For thousands of years groups of people were able to survive without possessing either clocks or calendars. Their members did not develop an individual conscience urging them constantly to take their bearings from a continuously advancing time. This does not mean that they lacked an individual conscience. As has been mentioned, people are so constituted that their chances of survival either as individuals or as groups are low if their natural potential for self-control and the regulation of momentary instinctive and affective impulses is not developed from childhood onwards in conformity to specific regulative patterns of behaviour and feeling.

What changes in the course of a civilizing process are, above all, the patterns of self-regulation and the way they are incorporated. In this respect the example of time-consciousness is helpful. The model of a civilizing process is not infrequently vulgarized today. For example, the central feature of the process is sometimes taken to be solely a continuous increase and reinforcement of self-restraint. This can give rise to misunderstandings. Shifts in this direction do occur. But this way of understanding the theory of civilization suggests that in simpler societies self-restraint is uniformly weak or absent, that affective or instinctual outbursts are equally strong in all areas of life. What characterizes the social habitus of people of simpler societies in reality is the uneven, often discontinuous character of the self-restraint. In certain life-situations it can be of an extent and severity far exceeding the level of self-control required in more developed societies. At the same time, in other situations it leaves the way open to a release of instincts or affects far more violent and spontaneous than the behaviour patterns that would be felt as tolerable in more developed societies. The property of the self-control patterns of people in simpler societies which is especially striking in comparison to those of people in highly developed industrial societies is their lack of uniformity. A powerful ritualization and formalization of behaviour with corresponding caution and self-restraint in some situations often goes hand-in-hand with an unbridled liberation of the affects in others. The swings of behaviour from one extreme to the other tolerated and even encouraged by society are greater. And the relation between external and self-imposed compulsion is different in less developed societies, if only because the pervasive sense customary in such societies of living in a world of spirits is a factor constantly affecting behaviour and feelings.

The dominant notion of living in a world of spirits is a common characteristic of stages in the development of knowledge at which certainty over the difference between animate beings and inanimate events has not emerged, and cannot yet do so. In the symbols of groups at such a stage this distinction is blurred. It is possible for everything that happens to have a will and intentions and to act in a similar way to human beings, the sun as much as a tree, a rock as much as a boat. From the perspective of later stages of society these notions can be recognized as collective or individual fantasies. Symbols of this kind have functions both of orientation and of regulation. But it is not easy to answer the question whether these are external or internal compulsions. People learn to restrain themselves or, as the case may be, to exert force on others, because they believe the spirits desire it. The idea of spirits can therefore have the functions of conscience. Groups which may not yet be in a position to control their instinctive or affective impulses on their own initiative are helped by the apparent external compulsion by fantasy figures which reinforce their self-control.

At all events, the intentions of spirits and other quasi-human fantasy beings are far less predictable than the regular and mechanical sequences that are attributed to many natural processes at later stages. At earlier stages the level of danger is higher, the level of personal security correspondingly lower. The formalization and ritualization of intercourse with spirits and human beings serve in this case to allay the uncertainty and fear to which this level of danger gives rise. For example, one often finds in simpler societies a very precise observance of rituals in social behaviour the significance of which includes a function as status symbols, as in greeting a visitor or the relations between sons and their father. The exact observance of such rituals is important here because the whole level of danger, including the threat to status, and the insecurity of people in their dealings with each other and with natural forces, is greater. In addition, loss of status in groups with low local mobility, once it occurs, is usually quite irreversible.

Perhaps it is only the comparison with simpler societies that makes one aware how much lower the level of danger and personal insecurity has become in developed societies, not least through the advanced control of natural forces and reduced dependence on their fluctuations. Accordingly, the pattern of self-restraint in people of more developed societies differs from that in simpler societies in a specific way. By and large, self-restraint at later stages follows a middle course; as compared to that of simpler societies it is more moderate and uniform, but also, like conformity to time, more inescapable. Like this conformity it extends fairly evenly into all areas of life. It may be somewhat relaxed in the most private relationships. But on the whole, social control resists excessive variations in individual self-control. We see once again that what characterizes the difference between earlier and later phases of a civilizing process is not the absence of self-restraint in the former and its presence in the latter. What changes is the relation between external and self-imposed constraints, and especially the pattern of self-restraint and the way it is incorporated.

But this also means that a mode of thought so widely established and highly respected that it is usually taken as self-evident, turns out nevertheless to be inadequate. People in more developed societies tend to regard the peculiar compulsions of their characters by which they differ for both good and ill from other people, as innate, part of their nature. Their compulsive self-control in conformity with tirelessly advancing clock and calendar time is a good example – one of many – showing that it is not only the genetically determined compulsions of their own nature but also constraints exerted by the social habitus conditioned by membership of a particular society, that play a decisive part in forming the individual character of a person. But to point this out is to take leave of an image of humankind predominant in the great canon of modern thinkers. From Descartes to the Existentialist philosophers of the twentieth century it has always been the society-less person who, whether in naturalistic or metaphysical attire, has played the central role. Often enough it is even a person who seems to exist without a world, independent of the natural universe. It is a curiously egocentric tradition, focused entirely on the self, the individual.

The study of time submitted here conveys a different picture. The isolated individual no longer stands at the centre. Nature is no longer a world of objects existing outside the individual; society no longer only a circle of others among whom the individual finds himself as if by chance. Calendar time illustrates in a simple way how the individual is embedded in a world in which there are many other people, a social world, and many other natural processes, a natural universe. By means of a calendar one can determine with great exactitude the point at which one entered the flow of social and natural processes. The recurring time-patterns of the days of the month show symbolically the return of one's day of birth, and the never-returning, ever-advancing year numbers of the time-scale operated by the society concerned, whether Islamic, Christian, Jewish or Japanese, provide the individual with a series of symbols with which he can determine exactly how often since his date of birth the sun has returned to a particular position standardized by society in its apparent movement in the sky or, in other words, how many years he has behind him. By means of calendar time the age of societies can also be determined, or the length of social processes and their stages and epochs. In recent times cosmologists have taken over the length of a physical sequence first standardized for social purposes as a year, to determine the duration or speed of other physical sequences. Not only the age of an individual person or the length of an individual lifespan, not only the 'age' of societies and the length of social processes, but the 'age' of the universe in which we live can be subjected to exact definition only in relation to event-sequences such as the solar year. This definition allows people to orient themselves better in their world, and perhaps even to gain better control of the dangers which threaten them.

That is not to say that this measurement of the duration of cosmic evolution and of other natural processes by the geocentric standard symbol of the solar year, which finally represents a certain number of apparent revolutions of the sun around the earth, does not correspond to reality. But it also indicates that it is only for beings like humans that there is any meaning or purpose in establishing such ordered relationships between processes like those of the universe or of light with the apparent revolutions of the sun. Only at the human stage do natural formations acquire the capacity for synthesis which enables them, with the aid of their social symbols, to imagine together and simultaneously the unfolding of the universe and the apparent circling of the sun about the earth. A long social development is needed before people evolve symbols for such complex ideas, without which they can neither communicate with each other about such ideas nor orient themselves according to them.

Time, as has been said, is a symbol of this kind of socially learned synthesis. One of the difficulties in investigating time is that people are as yet little aware of the nature and functioning of the symbols they have themselves developed and constantly use. They are therefore always in danger of losing themselves in the undergrowth of their own symbols. Time is one example. Human-made calendars and clockfaces bear witness to the symbolic character of time. But time has often appeared mysterious to human beings. Much remains to be done in elucidating the special nature of human symbols. To contribute to such elucidation is the object of this book. It therefore lacks, I must confess, any topicality. It is concerned neither with leisure time nor with working time. It hardly touches on questions of the day. Perhaps what is presented here can be taken as an example of the value of exploring, from time to time, layers of the communal life of people that are relatively untouched by issues of the day.

I have been able to say only as much on the nature of social symbols as is needed to steer the discussion of the human concept of time between the traditional philosophical alternatives, subjectivism and objectivism, nominalism and realism, and so to impart to people who are themselves affected by the controls issuing from the concept of time a clearer understanding of themselves and of the human situation.

One may perceive in reading this book that by means of the study of time a quite wide-ranging sociological theory is being developed further from a number of focal points. One of these focal points is a sociological theory of knowledge and perception[2] – sociological because the subject of knowledge is not taken here to be the individual person but the stream of many generations of people or, if you will, evolving humanity. The difficulty is that such a change in the field of study causes many familiar concepts to become useless or less significant. For example, changes in the direction discussed here take place at a level of social development at which the distinction between the 'history of knowledge' and the 'systematization of knowledge' breaks down. The learning process of humanity, which expresses itself among other things in a change in the measurement and experience of time by human groups, concerns neither the former nor the latter. And concepts like 'positivist' or 'transcendental' lose their meaning if one transfers one's analysis from the acquisition of knowledge by an individual to its acquisition by humanity, to the development, which has no beginning, of social symbols functioning as means of orientation.

If one makes this transition one finds that a central aspect of this process enacted by humankind is a specific change in the human attitude to the objects of knowledge, and undoubtedly not only to them. This change is expressed unequivocally in alterations in the structure and form of human orientation symbols. The receding of the question: 'Who is the originator of thunder and lightning?' and the advance of the question: 'What is the cause of thunder and lightning?', is an obvious example of this change. Within this theoretical structure the change has been characterized as a transformation in the relationship between human involvement and human detachment.[3] Changes in the balance between detachment and involvement can serve as means of sociological diagnosis in determining structural changes in orientation symbols, that is, in knowledge. They furnish the investigator with criteria for determining the direction of changes in the attitudes of people to each other, to themselves and to what we refer to, using a term which represents a relatively high level of detachment, as 'nature'. The primary perception of nature as a realm of spirits is thus characteristic of a stage of greater involvement, its perception as such, as nature, of a stage of a greater detachment, in the person experiencing it. But such shifts of balance are reversible. Even if, with regard to the perception and manipulation of natural sequences, detachment has become clearly dominant, the possibility cannot be ruled out that humanity might revert to a position in which the social symbols of a highly detached mode of speech, thought and knowledge are changed back into social symbols in which emotional involvement is dominant. Examples of such a change are to be found in the period of the gradual disintegration of the western Roman Empire. But limited shifts in this direction can be repeatedly observed even within periods of increasing detachment. In addition, the balance between detachment and involvement can be very different at the same time in different branches of knowledge. At present, the dominance of detachment is considerably greater and more secure in the knowledge and perception of natural processes than in the case of social processes. With the help of these and other criteria different stages of development, whether within one group or between various groups, are made comparable and the differences between them more clearly definable.

The development of the fund of human knowledge has an importance for humankind which is perhaps somewhat underestimated today. For this reason it is worthwhile to pay it rather more attention. In so doing one should not lose sight of the fact that the development of knowledge, whether in terms of greater detachment and reality-congruence or in terms of a greater involvement, in which symbols are invested with higher fantasy content, is inseparable from the direction of changes in the form and structure of the communal life of humans – for example, changes in the structure of conflicts and their resolution, in the social provision and distribution of nutrition, and in the distribution of all else that is necessary for survival. Like these other strands of development, that of the social fund of knowledge has survival functions for each group and its members: it is involved in their development not only passively but actively.

The case is similar with the development through civilization of the social habitus. Here too one can focus now on the direction of the overall development of survival units, such as tribes or states, and now on the direction of the development of these individual strands, of the civilisatory control of instincts and affects. If the focus is on the development of the latter, one quickly becomes aware of a problem which is highly characteristic of the indissoluble connection between the development of the communal life of people and that of the social personality structure of each individual among them, often with a marked time-lag. This is the problem of the relation of external compulsion to self-restraint.

Each person, to an extent, regulates himself. Each person is to a certain extent subject to compulsions arising from communal life with others, from the development and structure of his society, and finally from natural processes within himself or his society, for example, the natural compulsion to eat and drink, or those arising from wider nature such as heat and cold. People's scope for decision, their freedom, rests finally on the possibility that they have to control in a variety of ways the more or less flexible balance between various compelling authorities which, moreover, are constantly in a state of flux. Investigation into humanity will therefore remain empty-handed if it neglects the compulsions to which people are exposed or which they impose on themselves. Some researchers seem to believe they could open a way to freedom by ignoring such compulsions. But being ignored by investigators does not cause them not to exist. The relation of different kinds of compulsion to each other, the balances and constellations they form, differ widely at different stages of human development and between different social strata, so that people's scope for decision, their freedom, also differs between groups as between individuals.

The problem of the balance between individual self-restraint and external social compulsion first posed itself in the study on the civilizing process.[4] In the course of the present study on time it has emerged again. So, in a sense, the circle is closed. The three studies on the civilizing process, on the relation of involvement and detachment and on time, address related and often the same problems from different sides. Thus, what was elaborated previously with detailed empirical evidence on the direction of civilizing processes finds confirmation and further extension in what is revealed in the preface and the text of this book on the peculiar connection between external and internal constraint in the case of time. The social time-compulsion which has been largely converted into internal self-constraint proves to be a paradigmatic example of a type of civilisatory constraint frequently encountered in more developed societies. Members of these societies can often observe in themselves the internal compulsion to orient oneself in terms of time, whereas other forms of civilisatory self-constraint may be less easy to grasp as such.

The problem of social symbols has been pursued here only as far as seemed necessary for an understanding of the problem of time. Like a large number of other social symbols, that of time can fulfil several functions at once. The word and concept of time are together an example of a communicative symbol. A particular sound-pattern which can differ from society to society – 'time' in English-speaking societies, *temps* in French and *Zeit* in German – is associated for each member of the society with a learned memory pattern, often also called a 'meaning'. On the basis of this pre-agreement the sender of the sound-pattern 'time' can expect that the receiver, if he belongs to the same linguistic group, will associate the same memory pattern with this sound-pattern as he does himself. That is the secret of the communicative function of human symbols.

At the same time these symbols can also have the function of means of orientation. Physicists use the symbol 'time' in this way. But, beyond that, the face of every station clock shows that it is a socially institutionalized means of orientation. The mechanical apparatus of the station clock that is constantly moving in a specific way, transmits an institutionalized visual message to each person who is able to connect these sight patterns with the correct memory pattern. Finally, the orientation function of time is combined with a further function, that of a means of regulating human behaviour and feeling. The message of the station clock may cause a person to set himself in extremely rapid motion or, as the case may be, to adjourn for a lengthy wait to a nearby restaurant. Here the external and internal regulation of the individual person interlock in a peculiar way. The multi-functional character of time in more complex societies corresponds in this way to the extensiveness and diversity of its use.

Of this range of use I shall give, in conclusion, one more, perhaps somewhat unexpected, example. One might raise the question how the symbolic character of time is to be reconciled with the fact that time is also extensively used as a dimension of the natural universe. This is perhaps particularly surprising if one speaks, as has happened here, of a five-dimensional universe. In it time, as it appears, leads a strange double life. Everything that is perceptible, humans included, has a position in each of the four dimensions of space and time. But at the same time, time is being presented here as a symbol and so as a representative of the five-dimensional human world. What is the relation between these definitions of time, its definition both as a dimension of everything perceptible and as a social symbol which gradually evolves in the communal life of humans?

The answer, in a nutshell, is as follows: there are events which can be perceived as events in the stream of succession, that is, in time and space, without the perceiver taking into account the symbolic character of time and space. In this case one does not consciously consider that a learned synthesis of consciousness by order-creating people is needed in order to perceive perceptible processes as something taking place in time and space.

But if this is taken into account it is possible to climb one step higher on the spiral staircase of consciousness. Then, at the same time as the four-dimensional event, the fifth dimension too, that of the human beings who perceive and organize the event in time and space, comes into the observers' view. In this case, they see themselves as observers of the four-dimensional event standing, as it were, on the next-lower step of the staircase. In this way not only the four-dimensional event as such becomes visible, but also the symbolic character of the four dimensions as means of orientation for human beings – human beings who are capable of synthesis and so are in a position to have present at the same time in their imagination what takes place successively and so never exists simultaneously. Time, which on the preceding step was recognizable only as a dimension of nature, becomes recognizable, now that society is included in the field of view as a subject of knowledge, as a human-made symbol and, moreover, a symbol with high object-adequacy. That time takes on the character of a universal dimension is nothing other than a symbolic expression of the experience that everything which exists is part of an incessant sequence of events. Time is an expression of the fact that people try to define positions, the duration of intervals, the speed of changes and suchlike in this flow for the purpose of orientation. As a reply to the question raised previously, therefore, one needs a model of people who can observe and investigate from different storeys and so from different perspectives. If a four-dimensional universe is in question, people do not yet include themselves as observers and perceivers in their observations and perceptions. If one climbs to a higher step of knowledge and includes mankind as the subject of knowledge in one's knowledge, the symbolic character of the four dimensions also becomes recognizable.

Time: An Essay

This is an essay on time, but it is not concerned with time alone. One may notice soon enough that it also deals with a wider problem. For the perception of events which happen one after another as a 'sequence in time' presupposes the emergence within the world at large of beings, such as humans, who are capable of remembering distinctly what happened earlier and of seeing it in their minds' eyes as a single picture, together with what happened later and what is happening now. To perceive time, in other words, requires focusing-units (humans) capable of forming a mental picture in which events A, B, C, following one after another, are present together and yet, at the same time, are seen clearly as not having happened together; it requires beings with a specific potential for synthesis which is activated and patterned by experience. The potential for this kind of synthesis is a property peculiar to human beings; it is characteristic of their way of orienting themselves. Humans orient themselves less than any other creature we know by means of unlearned reactions and, more than any other creature, by perceptions which are patterned by learning, by previous experiences, not only of each human being individually, but also of long chains of human generations. This capacity for intergenerational learning, for handing on experiences of one human generation to another in the form of knowledge, is the basis of the gradual improvement and extension of their means of orientation over the centuries.

That which one today conceptualizes and experiences as 'time' is just that: a means of orientation. As such, it had to be evolved through experience in a long intergenerational process of learning. There is ample evidence to show that human beings did not always experience connections of events in the manner now symbolically represented by the concept 'time'. The potential for synthesis with which they are equipped had to be activated and patterned by experience and, specifically, by a long line of intergenerational experiences, before humans were capable of having the kind of mental picture of time-sequences which we now possess. In other words, humankind's experience of what is now called 'time' has changed in the past and continues to change today; it has changed and is changing, moreover, not in a random or historical manner, but in a structured and directional manner which can be explained. The task of this essay is to show, in broad outline, some aspects of the structure and direction of these changes and to indicate how one might set about explaining them. Basically, the salient points of this programme are simple enough.

1   From Descartes to Kant and beyond, the dominant hypothesis concerning time was based on the assumption that humans were endowed, as it were by nature, with specific ways of connecting events, of which that of time was one. It was assumed, in other words, that the synthesis of events in the form of time-sequences patterned humankind's perception prior to any experience and was, therefore, neither dependent on any knowledge available in their society nor acquired through learning. The assumption of such a 'synthesis *a priori*' implied that humans possessed not only a *general* capacity for establishing connections, but also a compulsory capacity for making

specific connections and for forming corresponding concepts such as 'time', 'space', 'substance', 'natural laws', 'mechanical causation' and others, which were thus made to appear as unlearned and unchanging.

I am going to show that this hypothesis is untenable. Humans possess, as part of their natural endowment, a general potential for synthesis, that is, for connecting events, but all the specific connections which they establish and the corresponding concepts used by them in their communications and reflections are the result of learning and experience, not simply of each individual human being, but of a very long line of human generations handing on knowledge and learning from one to the other: an individual life is far too short for the learning process needed in order to acquire the knowledge of specific connections such as those represented by concepts like 'cause', 'time' and others of equal universality. The philosophical view that people connect events in the form of 'time', as it were, automatically and without any learning, by means of a 'synthesis *a priori*' as a gift of their native reason, was due partly to the limited evidence available to – or used by – Descartes, Kant and others who followed in their footsteps, and partly to their concept of experience: when speaking of experience, they had in mind only the experience of a single person conceived as a totally autonomous entity and not the experience and knowledge of humankind as it grows over the centuries.

2 The idea that people have always experienced the sequences of events which one now experiences as time-sequences in the manner which predominates today – namely, as an even, uniform and continuous flow – runs counter to evidence we have from past ages as well as from our own. Einstein's correction of the Newtonian time-concept is a contemporary example of the way in which the concept of time can change. Einstein

made it clear that the Newtonian hypothesis of time as a unitary and uniform continuum throughout the whole of the physical universe was untenable. If one takes the trouble to look at earlier stages in the development of human societies one can find ample evidence of corresponding changes in people's experience and conceptualization of what we now call 'time'. As used today, 'time' is a concept at a high level of generalization and synthesis[1] which presupposes a very large social fund of knowledge about ways of measuring time-sequences and about their regularities. At an earlier stage people evidently could not have had this knowledge – not because they were less 'intelligent' but because this knowledge, by its very nature, required a long time to develop.

Among the earliest time-meters were the movements of the sun, moon and stars. We have a very clear picture of the connections and regularities of these movements; our ancestors had not. If one goes back far enough one encounters stages where people were not yet in a position to link the varied and complex movements of the heavenly bodies to each other in the form of a relatively well-integrated unitary picture. They experienced a great mass of single items without clear connections or, at the most, rather unstable fantasy connections. If one has no firm yardstick for timing events one cannot have a concept of time akin to ours. Furthermore, at an earlier stage people communicated with each other – and thought – in what are often called today more 'concrete' terms. As no concept can be, properly speaking, regarded as 'concrete', it would probably be more correct to speak of 'particularizing' or 'low-level' abstractions. There were stages at which people used the concept 'sleep' where we would speak of 'night', the concept 'moon' where we would speak of 'month', the concept 'harvest' or 'produce of the year' where we would speak of 'year'. One of the difficulties one encounters in an essay on time is the absence of a developmental theory of abstraction. The changes just mentioned from particularizing to generalizing abstractions are among the most significant developmental changes one encounters in this context, but there will not be enough room here to enlarge upon it. Moreover, specific time-units such as 'day', 'month', 'year', etc., which now flow smoothly into each other in accordance with our calendar and with other time regulations, did not always do so in the past. In fact, it is the development of time-reckoning in social life and of a relatively well-integrated grid of time-regulators such as continuous clocks, continuous yearly calendars and era time-scales girding the centuries (we live in the 'twentieth century Anno Domini') which is an indispensable condition of the experience of time as an even, uniform flow. Where the former is lacking the latter is lacking too.

3 A model of the development of time-concepts enables one to see more clearly the growth of the relative autonomy of society within nature. There are earlier stages where the social enclave within nature built by humans is small. The interdependence between these enclaves and what we now call, in our self-centred manner, 'our environment' is obvious and direct. The balance of power between human groups forming these enclaves and non-human nature is weighted more strongly in favour of the latter; and the timing of social events is largely dependent on observations about recurrent, natural and non-human events. As the human enclaves grow in size as well as in relative autonomy – in the course of processes such as urbanization, commercialization and mechanization – their dependence on human-made devices with the function of time-meters and time-regulators grows and that on non-human, natural time-meters such as the movements of the moon, the changing seasons, the coming and going of the tides, diminishes. In the highly urbanized and

industrialized societies of our age the relationship between the changing calendar-units and the changing seasons, while not entirely disappearing, becomes more indirect and tenuous and, in some cases, such as that between the month and the movements of the moon, it has more or less vanished. People live, to a much greater extent, within a world of symbols of their own making. The relative autonomy of their social enclaves, without ever becoming absolute, has vastly increased. One may add that the development in that direction is anything but irreversible, neither is it straight; there are many reversals, detours and zig-zag movements. Moreover, given the multifarious nature of humankind development, one can encounter developmental sequences in the direction towards autonomy, recurring at very different dates within the era time-scale used today. Structurally equivalent stages, e.g. those with hardly any or with no human-made timing devices may be encountered before, as well as after, the nineteenth century of the presently used era time-scale. A reminder of the fact that the autonomy of people's social enclaves, though it can increase, is always relative, may help to counter a grossly misleading habit of thinking which has grown among us. We are apt to think and to speak in terms which suggest that 'society' and 'nature', 'subject' and 'object', exist independently of each other. This is a fallacy which is hard to combat without a long-term perspective.

4   In discussions about the problem of time one is liable to be misled by the substantival form of the term. I have pointed out elsewhere[2] that the convention of speaking and thinking in terms of reifying substantives can gravely obstruct one's comprehension of the nexus of events. It is reminiscent of the tendency of the ancients, which has by no means entirely disappeared today, to personify abstractions. Just actions became the goddess Justitia. There are

plentiful examples of the pressure which a socially standardized language puts on the individual speaker to use reifying substantives. Take such sentences as: 'The wind is blowing' or 'The river is flowing' – are not the wind and blowing, the river and flowing, identical? Is there a wind that does not blow, a river that does not flow?

The case is similar with the concept of time. Here, too, Western linguistic tradition has transformed an activity into a kind of object. But in the case of time the activity represented by the verbal form of the word is not a natural but a social activity. Perhaps it is easier to recognize this fact if one uses a language such as English which offers those who speak it a verbal as well as the substantival form. The Oxford Dictionary quotes the sentence from Bacon: 'There is surely no greater Wisdom than to time the Beginnings and Onsets of things.' Or from more recent times, 'to time' is defined as: 'To adjust the parts of (a mechanism) so that a succession of movements or operations takes place at the required intervals and in the desired sequence' (1895). The verbal form 'to time' makes it more immediately understandable that the reifying character of the substantival form, 'time', disguises the instrumental character of the activity of timing. It obscures the fact that the activity of timing, e.g. by looking at one's watch, has the function of relating to each other the positions of events in the successive order of two or more change-continua.

Languages which lack a verbal form corresponding to the English 'to time' offer to members of the societies where these languages are spoken only expressions such as 'to determine the time' or 'to measure the time'. That still suggests the existence of a thing called time which one can measure or determine. Linguistic habits are therefore misleading for reflection. They constantly reinforce the myth of time as something which in some sense exists and as such can be determined or measured even if it cannot be perceived by the senses. On this peculiar mode of existence of time one can philosophize tirelessly, as has indeed been done over the centuries. One can entertain oneself and others with speculations on the secret of time as a master of mystery, although there actually is no mystery.

It was Einstein who finally set the seal on the discovery that time was a form of relationship and not, as Newton believed, an objective flow, a part of creation such as rivers and mountains which, although invisible, was like them independent of the people who do the timing. But even Einstein did not probe deeply enough. He too did not entirely escape the pressure of word-fetishism, and in his own way gave new sustenance to the myth of reified time, for example, by maintaining that under certain circumstances time could contract or expand. He discussed the problems of time only within the limited terms of reference of a physicist. But a critical examination of the concept of time requires an understanding of the relation between physical time and social time or, in other words, between timing in the context of 'nature' and in that of 'society'. But this was not his task, nor did it fall within the competence of a physicist.

5 The steady expansion of human societies within the non-human, the 'earthly' sector of the universe, as mentioned before, has led to a mode of discourse which gives the impression that 'society' and 'nature' exist in separate compartments. The divergent development of natural and of social sciences has reinforced this impression. The problem of time, however, is one which we cannot hope to solve so long as physical and social time are examined independently of each other. If one translates 'time' into its verbal form and examines the problem of timing, one can see at once that the timing of social events and of physical events cannot be entirely separated. With the development of human-made time-meters, the relative autonomy of social timing in relation to the timing of non-human physical

events increased; their connection became more indirect but it was never broken – it is, in fact, unbreakable. For a long time, however, the social requirements of people provided the impetus for time-measurements of the 'heavenly bodies'. It is not difficult to show how much the development of the latter, in spite of all reciprocity, was and remained dependent on that of the former.

Unless one keeps in mind the unbreakable relationship between the physical and the social levels of the universe – unless one learns, in other words, to perceive human societies as emerging and developing *within* the larger non-human universe – one is unable to attack one of the most crucial aspects of the problem of time. One can present it briefly in the following manner: in the context of physics and, thus, of the ruling tradition of philosophy, 'time' appears as a concept at a very high level of abstraction, whilst in the practice of human societies 'time' is a regulatory device with a very strong compelling force, as one can readily see if one is late for an important appointment. The conventional tendency to explore 'nature' and 'society' and, therefore, the physical and the sociological problems of 'time' as if they were completely independent of each other thus gives rise to a seemingly paradoxical problem which, as a rule, is tacitly swept under the carpet in discussions about 'time': how is it possible that something which appears, in general reflections, as a high-level abstraction, can exercise a very strong compulsion on people? To date, enquiries into the sociology of time are almost non-existent. The fact that time is still largely discussed in the traditional philosophical manner, even by sociologists, has something to do with it. Another reason is that studies in the sociology of time cannot be very productive as long as they are tied to a short-term perspective. They can come into their own only within the framework of a developmental and comparative approach guided by a long-term perspective.

6   Many familiar figures of speech, as has been noted, give the impression that time is a physical object. Simply by speaking of 'measuring' time one makes it appear as if time is actually a physical object like a mountain or a river, the dimensions of which can be measured. Or consider the expression 'in the course of time'; it almost suggests that people, or perhaps the whole world, were swimming in a river of time. In this case as in others the substantival form of the concept of time undoubtedly contributes much to the illusion that time is a kind of thing existing 'in time and space'. The verbal form makes it easier to avoid this illusion. It makes it clear that time-measuring or synchronization is a human activity with quite specific objectives, not merely a relationship but a capacity for establishing relationships. The question is: Who in this case relates what, and to what end?

The first step towards an answer is relatively simple: the word 'time', one might say, is a symbol of a relationship that a human group, that is, a group of beings biologically endowed with the capacity for memory and synthesis, establishes between two or more continua of changes, one of which is used by it as a frame of reference or standard of measurement for the other (or others).

The ebb and flow of the tide, or the rising and setting of the sun and moon can serve as socially standardized continua of changes of this kind. If these crude natural sequences are found too imprecise for their purposes, people can on their own initiative establish more exact and reliable sequences as a standard for other sequences. Clocks are precisely this; they are nothing other than human-made physical continua of change which, in certain societies, are standardized as a framework of reference and a measure for other social or physical continua of changes.[3]

To relate different continua of changes to each other as 'time' is therefore to link at least three continua: the people who connect, and two or more continua of changes, one of which takes on, within particular societies, the function of a standard continuum, a framework of reference, for the other. Even if an individual member of such a group uses *himself* as the frame of reference, as in the case in which one uses one's own life as the standard continuum for timing other events, the relationship is functionally three-polar: there is oneself as the person who integrates and times; there is oneself perceived as a continuum of changes from birth to death and, in that capacity, used as a standard continuum; and there is the host of other changes which one measures in terms of the span of one's own life – of oneself as a continuum of changes.

To avoid misunderstanding, I may add that it is only in quite highly individualized societies that a person is capable of timing events in terms of his own life as a standard continuum. In these societies, not only does each person stand out more distinctly from all others as a uniquely patterned individual, but each person is also able to time his own life very exactly as a continuum of changes in terms of another socially agreed standard continuum, such as that of the sequence of calendar years. It is at a somewhat later stage in the development of human societies that men encounter the problem and are capable of working out a relatively well-fitting time-reckoning in terms of an era, such as the Greek dating of events in terms of the sequence of olympiads, the Roman counting of years '*ab urbe condita*', or the corresponding Jewish and Christian era-timing.

In societies where no long-term era-calendar exists as a man-made standard continuum such as these, people – understandably – are unable to say when they were born or how old they are except in terms of single events such as 'when the great storm came' or 'just before the old chief was killed'. They have, in this case, no social standard continuum as frame of reference for the continuum of changes which each of them is himself. Thus, whenever one uses, in societies such as ours, one's own individual life as a frame of reference for timing other changes, one also uses implicitly a socially evolved continuum of changes, the era-reckoning of calendar years, as standard for a continuum which oneself is.

Time-relations, as one can see, are many-layered relationships of considerable complexity. One may experience some difficulty in distancing oneself from the homely metaphors which make 'time' appear as a thing, or the widely accepted notion that time can be used as a kind of football for philosophical fantasies, since nothing definite can be said about it. But in the long run one will find it more rewarding to adopt a mode of thinking which reveals 'time' to be a conceptual symbol of a gradually advancing synthesis, a setting-up of fairly complex relationships between continua of changes of various kinds.

In its most elementary form, 'timing' means determining whether a change – which may be recurrent or non-recurrent – happens before, after, or simultaneously with another. If it is a sequence of changes it may mean answering questions such as that of the length of the interval between them in terms of a socially agreed timing standard, such as the interval between two harvests, or between one new moon and another. On a more differentiated level, one may time the interval between the start and finish of a run, a reign, a human life or between what we call 'antiquity' and 'modern times', in terms of a socially agreed standard continuum of changes. How far human groups can 'time' events and can, thus, experience them in terms of 'time' depends on the extent to which they encounter in their social practice problems which require timing and on that to which their social organization and their knowledge enable them to use one set of changes as a standard for another.

7 'When shall we do it?' This is the primary question in response to which people set out on the adventure of timing. The point of departure, that which one endeavours to time, is in the first place always one's own activities and, in the earlier stages, primarily one's own activities as a group. However, in a sense, people 'time' their activities at a stage where they are not yet confronted with problems which find expression in explicit and articulate when-questions. Timing in these stages is passive; it is hardly reflected on and conceptualized as such and, to some extent, passive timing remains with us. Thus one may time one's activities more or less in accordance with the prompting of one's own animalic urges: one may eat when hungry and go to sleep when tired. In our type of society these more animalic cycles are regulated and patterned in accordance with a differentiated social organization which compels people, up to a point, to discipline their physiological clockwork in terms of a social clockwork. This is far less the case in relatively simple societies; there the regulation and patterning of the physiological clockwork (as far as one can call it 'regulation') depends more directly on the extent to which non-human nature, or, in some cases, other humans on whom one can prey, allow or withhold the satisfaction of one's needs. In these societies men may go hunting when they feel hungry and may stop exerting themselves when they have eaten enough. Or, on a slightly higher level in the sequence of passive timing, they may go to sleep when it gets dark and get up when the sun rises. Conceptualization may be in terms of 'sleep' rather than in terms of 'night'. There are, thus, stages in the development of human societies where men have hardly any social timing problems which require an active synchronization of their own communal activities with other changes in the universe.

The scenario becomes markedly different when people engage in actively producing their food. Agriculture, the utilization of domesticated plants, is a good example. At this stage, problems of active timing (in addition to passive timing) and, thus, of active, social and personal control, become more pronounced. For, by controlling and utilizing vegetation, men become subject to a previously unknown discipline imposed on them by the requirements of the domesticated vegetation on which they have become dependent for their food supply. An example taken from the nineteenth-century history of a small African people may illustrate the point. It indicates briefly one of the problems which induce early agriculturalists to evolve forms of active timing:

Another work that the ... priest was in duty bound to do, was to watch the seasons in order to enable him to announce or proclaim to the public in general the time for solving wheat, and also that concerning the celebration of their festivals.

On account of the former he had to go to the top of one of the sentinels facing east to watch the sunrise every morning. It is said that there is a table mountain ... on the east ... and when the sun is seen to rise exactly behind this Table-Mountain, then the first rain in that week was considered enough for sowing. The next morning after the rain the priest would give an alarm that would be re-echoed throughout the mountain home. After this you would see the farmers and their families running down the mountain with their hoes and baskets to join or share in the labour.   This is the alarm:

> Throw away poverty or
> Famine is over
> Although never is it said
> Day or night;
> I am to say it now
> In order to be sent away
> To the land of suffering.

You would hear people sounding this alarm when sowing was going on, but after, he or she who dared (sc. to say this verse, to use the magical formula for sending Famine to the land of suffering) would be severely punished or, worst of all, sent away into slavery.

Now in order to enable him also to tell the people exactly of their mid-yearly celebrations of their festivals, the Priest had to go to another rock facing West, at the sight of every New Moon to make a mark there with a stone or put a cowry into a pot set there for the purpose. Nobody had to touch this besides the priest or his assistant...[4]

The quotation illustrates most vividly the time-experience of people at a relatively early agricultural stage. It is determined by their practical social needs. It is, in that sense, strictly self-centred – not in terms of an individual self, but in terms of the group which experiences a timing problem. The priest observes the passage of sun and moon not because he is interested in astronomy but because these changing lights in the sky, and through them perhaps some unseen powers, tell him *when* his people should begin their sowing and *when* they should celebrate their cult festivals with rituals and sacrifices and, perhaps, with singing and dancing, so as to enlist the help of their gods in the production of food and in warding off all kinds of danger. At this stage, food production and cult activities are still closely related. Together, they are among the first social foci which present people with problems of active timing. Passive timing requires no decision; active timing does. The crucial point is the co-ordination of the continuous round of social practices with the continuous round of changes in non-human nature. For example, last year's harvest stored for the lean months may be almost exhausted. Although perhaps supplemented by meat from animals killed by hunters or by wild roots collected here and there, for a full larder one would have to wait until the next harvest. Given the, for people, unmanageable cyclical change of the seasons and the relatively more manageable rhythm of plant growth, the question one has to decide is: when should one start sowing? In the West African setting, that means: when will the dry season give way to the rainy season? Is the rain that has fallen its harbinger, or is it a false start? Now the sun, through the mouth of the priest, has given the answer; and the people rejoice. They themselves are probably not very concerned with the question as to how their priest has arrived at an answer. They have not yet a sense of 'time' in the abstract, as something that is passing. They are concerned with their immediate problems, such as that of their decreasing stores. There are certainly societies at a comparable level of development at which such experiences have not yet crystallized into regulating concepts at the comparatively high level of abstraction, such as 'month', 'year' or, for that matter, 'time'. Their concepts are much more closely linked to the recurrent cycles of their tangible needs, of transient satisfactions giving way to renewed needs and the quest for further satisfactions in an unending round.

Timing at this stage has still very much more the character of taking omens than that of looking at an impersonal clockwork in the sky. Only gradually will it take its place and its meaning somewhere between these two poles. Moreover, abstract counting by means of numerals alone is still difficult or entirely absent. Hence the priest tries to remember how often a new moon is seen in the sky after the hot dry winds appear to have come to an end, by putting a cowrie shell in a pot every time he observes a new moon passing a particular spot. The size of the heap of cowrie shells tells him roughly whether the right time for sowing has come. It may be hard to imagine, today, human life at a stage where the knowledge of abstract counting in our manner has not been evolved – as hard as it is to imagine a stage at which people have not yet evolved the continuous timing and dating techniques which are the condition of our experience of 'time' as a continuous flow. But it is well worth making the effort.

8   This episode from the history of a small hill-people shows many structural characteristics of relevance to a sociology of time. In sociological terms, time has a co-ordinating and integrating function. During earlier stages, the social function of co-ordination and integration is usually exercised by certain central figures, such as priests or kings. Co-ordination through knowledge of the 'right time' for doing things, in particular, is for a long time the special social function of priests. One can see it here. Priests are freed from the labour of growing their own food. They have more time for observing the movements of the changing lights in the sky. Already in this small African village-state the priest possesses, by virtue of his secret knowledge about finding the 'right time', sufficient power and authority to decide for the members of his society when the communal food-producing activities or, alternatively, the seasonal cult-actions should begin.

This is not an isolated case. Almost everywhere in the long development of human societies priests were the first specialists in active timing. When larger and more complex state-societies developed, priests usually shared the social function of determining when certain social activities should be undertaken – in an often uneasy partnership – with secular state authorities. When the struggles for supremacy between priests and kings went in favour of the latter, the setting of time, like the coining of money, became one of the monopolies of the state.[5] Yet even then priests remained for a long time specialists in timing procedures. In Assyria, for instance, priestly observers of the sky had to inform the king when a new moon had been actually seen by them in their special 'observatories'; for the more abstract anticipatory calculation of the moon's circuit around the earth was still beyond people's reach. In Athens a special priestly functionary, the Hieromnemon, an elected member of the highest ruling body, was concerned with the redaction of the new calendar from year to year (and Aristophanes probably got a big laugh from his audience by asking the new Hieromnemon to see to it that the days of the social calendar he had to fix were better synchronized with the visible cycles of the moon than were those of his predecessors). Caesar consulted the Pontifex Maminus when he wanted to improve the old calendar; it had got out of tune with the observable movements of the heavenly bodies and he evidently felt that it was a ruler's task to provide a well-functioning timing and dating framework for all civic activities.

Not surprisingly, the need for an orderly and unified time-reckoning varied in accordance with the growth and decline of state-units, with the size and the degree of integration of their peoples and territories and the corresponding degree of differentiation and length of their commercial and industrial ties. The juridical institutions of states required unified time-meters commensurate with the complexity and diversity of the cases which came before them. With growing urbanization and commercialization, the problem of synchronizing the growing number of human activities and of having a smooth-running continuous time-grid as a common frame of reference for all human activities became more urgent. It was one of the functions of central authorities, whether secular or clerical, to provide it and to see that it functioned. The orderly, recurrent payment of taxes, interest, wages and the fulfilment of many other contracts and obligations depend on it; and so did the many holy days – or holidays – when people rested from their labours. One can see very clearly how, under the pressure of these needs, church and state authorities, the holders of the monopoly of timing, attempted to cope with that task. For a long time, for instance, there were, even within one and the same state, traditional local diversities with regard to the beginning of a year and, thus, to its end. As far as one can see, it was

Charles IX, king of France, who, after some discussion, decided in 1563 to impose on French society a uniform date for the beginning of the year, setting it at 1 January. His edict, put into practice in 1566, broke with a more or less official tradition which linked the beginning of the year to the Easter festival. Accordingly, the year 1566, beginning on 14 April and ending on 31 December, had only eight months and seventeen days. Thus the months of September, October, November and December which, in accordance with a Roman tradition of begining the year in March and with the corresponding meaning of their names, had been the seventh, eight, ninth and tenth months respectively, now became somewhat incongruously the ninth, tenth, eleventh and twelfth months. At the time this change aroused strong opposition. Now one hardly notices the incongruity. Calendars as a social institution have a regulating social function. Now one takes it for granted that 1 January really is the beginning of the year. One does not see clearly that a year has a social function and a social reality related to, but distant from, a natural reality; one is apt to perceive it simply as something established by nature. Again, Pope Gregory XIII decided to revise the Julian calendar because, in the 400 years that had elapsed since it had been established, the spring equinox which determined the Easter festival had slowly receded from 21 March to 11 March. A Papal Bull suppressed ten days of the year 1582 and decreed that the day following 4 October was not 5 but 15 October. The Gregorian reform of Julius Caesar's reform of the old Roman calendar was the last attempt, so far, to provide a calendar system for a social year which, over the centuries, did not diverge too much from the 'natural year', that is, from the time in which the sun – in relation to men as observers and centres of reference – returns to a point in the sky which has been singled out by them as a point of departure.

As an example of developmental continuities persisting in spite of a good many political and other discontinuities – of what has been called before a 'continuum of changes' – the evolution of a timing framework in the form of a calendar is not uninstructive. It can serve as a reminder that what we call 'time' is an often rather complex network of relationships and that timing is essentially a synthesis, an integrating activity. In this case men use initially, as a standard for the relatively fast-moving continuum of social changes, the natural continuum of changes in the sky which is, by comparison, changing so slowly that, in relation to themselves, it does not appear to change at all. By fixing, more or less arbitrarily, a certain position of the sun, perhaps in relation to other stars, as the starting- and finishing-point of a social timing-unit – a year – they are able to establish a frame of reference for the synchronizing of human activities. The slowness with which people succeeded, over the centuries, in working out a calendar time-scale well-tailored in relation to the physical continuum and capable of providing an articulated, unified synchronizing standard for human beings integrated in the form of states and, now, beyond these, for a global network of states, shows how difficult this task was.

9   It was, apparently, even more difficult for people to establish an era time-scale for sequences of hundreds or thousands of years which enabled the living generation to determine with accuracy their own position in the sequence of generations. The working out of this type of *non-recurrent* time-scale raises particular problems. An early form is the counting of a sequence of passing years in terms of a sequence of rulers and the sequence of years in each reign. One of the widely used era time-scales of our time is that which counts centuries and years in the terms:

before and after the birth of Christ. To establish a measuring rod of this type for long non-recurrent time-sequences was possible only when social units such as states or churches had the character of a long-lasting continuum of changes within which living groups – usually ruling groups – found it necessary for the functioning of their institutions to keep alive the memory of the continuity of these constitutions in a precise and articulate manner. In antiquity the longest and best-known era time-scale was that which counted the non-recurrent sequence of years from the reign of a Babylonian ruler Nabonassaros. The time-reckoning in terms of consecutive reigns and their years, established first for purposes of state in the Chaldaic Babylonian tradition, made it possible to put on record, in a purely descriptive manner, the distance between unusual events in the sky, such as eclipses of the moon, in terms of the number of years that lay between them. Later, Ptolemy used this Babylonian era time-scale, the oldest and longest available to him within his knowledge continuum, for the construction of his model of the physical universe. His case illuminates the intertwining of the development of social and physical time-meters. Today it is often taken for granted by philosophers and, perhaps, by physicists that 'time flows in one direction, and the flow of time cannot be reversed', although Einstein's theory, while maintaining the serial order of time, questioned its uni-directional character. It is hard to imagine that physicists could have developed the concept of a unidirectional and irreversible flow of time within their sphere without the slow and difficult emergence of social time-scales, with the help of which one could accurately determine the non-recurrent, continuous sequence of years, centuries and millennia. The emergence of long-lasting and relatively stable state-units, in other words, was a condition of the experience of time as a uni-directional flow.

If one extends, in imagination, the developmental trajectory of timing backwards to the conditions of early agricultural societies, such as that mentioned before, one can see that very clearly and why neither elaborate yearly calendars related to recurrent events nor long-lasting era time-scales (the condition of an experience of 'time' as a continuous, irreversible flow) have at that stage come into existence. Problems which require a precisely subdivided timing framework for social activities divided into time-units such as 'months' or 'years' have not yet arisen or, if they have, are still difficult to solve; and problems concerning dozens or hundreds of years are, at most, dimly perceived in terms of a line of ancestors or entirely beyond one's horizon. A priest, as one saw, determines the 'right time' not yet by relating the movements of the sun to the 'fixed' stars but in relation to some earthly landmarks; and he tries to discover it, not yet by relating the activities of his people to a continuous calendar time-scale extending over a whole year, but only by relating the 'right time' in each instance to a particular event, such as sowing or organizing a feast for the gods.

10 In order to see in better perspective the whole trajectory of the development of timing and of standards for timing, one might go back even further. There is, indeed, no end to it – or, rather, no beginning. One can only find a mooring in the flowing continuum of evolutionary changes, many of which are still unknown or ill-understood, by constructing as a working hypothesis the scenario of an imaginary beginning. Imagine, then, a human group whose members have biologically exactly the same potential as we have for communicating with each other by means of common signals which are not inborn, which are symbolic representations of everything they experience – representations that can be learned, improved and handed down from one generation to another. They

themselves, however, have not been handed down any symbols of this kind from previous generations. The scenario is that of a group of humans to whom no knowledge and, therefore, no concepts have been transmitted from their ancestors. It represents a Rubicon model: the biological breakthrough to the new level at which organisms are able to make their own signals as the principal means of communication with each other has been achieved; a fifth dimension has emerged from within the, so far, four-dimensional universe. The specifically human adventure of creating a social universe where one communicates through human-made symbols capable of being learned, improved and extended has begun. But as yet people have no symbolic signs of this kind; they have not learned from their parents any conceptual means of connecting events – which means, in effect, that they have initially no means of distinguishing events; no 'objects', in our sense of the word, are known to them. They have to start from scratch. The telescoping into an instant situation of a process which took millions of years and which one would otherwise have to conceptualize by means of graduating terms, makes it possible to leave aside for the moment all the complex problems inherent in that long process of socio-biological changes and to concentrate on problems of the process of socio-symbolic changes within a biological setting that is unchanging, or changing so slowly compared with the continuum of social changes that one can neglect it here. Also, it may put too heavy a burden on one's imagination to ask how humans in that condition might experience their world. Perhaps it would be easier to ask which of the connections between events which one now establishes as a matter of course could be established by people without antecedent knowledge. Is it possible, for instance, to imagine that, starting from zero, they could work out – even within the life-span of a few generations – communicable signals, such as words, harbouring any of those connecting concepts at the highest level of abstraction whose possession has greatly exercised the imagination of philosophers – concepts such as 'cause and effect', 'nature and natural laws', 'substance' or 'time and space'?

Members of the initially conceptless group have the full biological potential for synthesis which men have today; they have the same capacity for connecting perceived events, but no knowledge as to how to connect them. All symbols of specific connections have yet to be worked out. They have the same faculty for learning, for giving each other learned signals symbolically representing all sorts of experiences, in addition to the more residual unlearned signals such as laughing, crying, sighing and other 'expressive' bodily movements with an inborn base which play an increasing part as means of communication as one moves backwards along the evolutionary ladder. However, they have only the faculty, there is no one from whom they can learn. Biologically the Rubicon has been crossed. They are human in the full sense of the word, but they start with a conceptual tabula rasa. Any knowledge they can acquire, any learned signals they may evolve – that is the assumption – they have to work out on their own through their own experiences. Is it possible to imagine that they would automatically experience the connection of events in terms of time-sequences, recurrent or non-recurrent? Is it possible to postulate, either, that, driven by an *a priori* condition of their 'reason', they would automatically experience everything in terms of time-sequences without even having worked out any time-meters, or that they could immediately set about working out time-meters on the basis of their experiences here and now? Or, if not here and now, how far could they get in one generation?

11 The term 'tabula rasa' is used here not entirely by accident. It played a part in the long-drawn-out discussion

about the question as to whether or not people possess inborn ideas. As the experience and the concept of time have often been treated as if 'time' were a kind of inborn idea, reflections on the scenario of a human group starting without antecedent knowledge and, thus, without knowledge of the concept of time, may be of some use here. Notions of specific inborn ideas are again and again smuggled surreptitiously, by the back door, into discussions about human 'language' and 'reason'. The labels may vary greatly; they range from terms like '*a priori*' or 'the unchanging conditions of human experience' to such expressions as 'deep structure of language' and 'eternal laws of reason or of logic'. Their exponents are frequently themselves unaware that they are referring to innate biological structures of humankind – since they present all the alleged specific structures of thought or language as unlearned properties common to all people irrespective of social learning. But what other possibility is there? One cannot have one's cake and eat it; nor can one say that an idea, a concept is *a priori* – an immutable property of human 'existence' or understanding, an eternal category prior to all human experience, in short, unlearned – without saying that it is inborn.

The thought-experiment of an initially conceptless human group may help to bring the issue into the open. It illuminates the hypothetical question: how would people, biologically equipped as they are, experience the world if no knowledge, and especially no knowledge of concepts, had been handed down to them as a result of a long antecedent intergenerational process with its continuous exchange, confrontation and blending of experience and conceptualization? Could they instantaneously connect events in accordance with the concepts of time or mechanical causality?

One can see at once the difference between this and the familiar Cartesian scenario. There, a seemingly timeless individual Ego meditates about his own 'reason' in complete isolation from the rest of the world. The meditating person tries to empty his 'mind' of all experiences, of all learned concepts, because they may be illusions. He is thus left with an intellectual tabula rasa in the sense that everything acquired through experience, all knowledge and, thus, all concepts learned from others, have disappeared. 'Reason', on that assumption, is a kind of unchangeable vessel which can be emptied of its changeable contents and is absolutely identical in all human beings. A long line of philosophers have followed in Descartes' footsteps; they have taken up and developed this scenario. In doing so they do not distinguish clearly enough between people's *universal potential* for synthesis, their capacity for making connections, and the postulation that 'a person' is endowed by nature and, thus, without learning, with the capacity for connecting events in highly specific ways, with innate concepts or ideas, such as 'cause', 'substance' or 'time'. These are presented as symptoms of an innate synthesis which compels people to connect events in this highly specific manner, independently of any knowledge or experience. They are thus made to appear as a predetermined and unchanging condition of every human experience throughout the ages.

The unsatisfactory nature of this scenario comes to light if one considers what a person is supposed to do in this case: he is supposed to penetrate in his meditation, all on his own, to a layer of his own intellect believed, in accordance with an unexamined dogma of his time, to be unlearned and independent of his own or anyone's experience. In trying to do so he deploys an immense arsenal of learned knowledge, including learned concepts. What he encounters in his descent into the transcendental depths of his own thought and what he brings to light is, in other words, the very conceptual apparatus that has been passed on to him by others, that he uses on his 'journey into the interior'. That is, he interprets as an unlearned property of his own thinking, and of that of all others, concepts which are part of the established repertoire of the language and knowledge of his time – but certainly not of all times – a repertoire which has emerged in this form from the practical and theoretical endeavours of a long chain of generations.

Thus Kant, as a representative of his stage in the long journey of humankind had learned to use the concept of 'time' in the sense it had acquired at that stage in conjunction, above all, with the progress in physics and the technical sciences, and then discovered – miraculously! – the very same concept within himself as an unlearned form of his own experience and that of 'humankind in general'. From his own subjective experience, Kant prematurely concluded that his experience and concept of time must be an unalterable condition of all human experience, without pausing to reflect that this hypothesis could be tested, that one could investigate whether, in fact, people everywhere and always have or had a time-concept at the same level of synthesis, which he declared to be a permanent condition of all human experience.[6]

Or take Descartes: having argued his case in the highly developed philosophical language of his age, he summed up his findings in the famous Latin tag 'cogito ergo sum', implying that everything outside himself might be an illusion and its existence doubted – only his own reasoning and, thus, his own existence could not be doubted. Yet all this was argued in languages such as Latin and French and thought out with the help of the tradition of knowledge handed on to Descartes together with these languages. He thus derived from what he had learned from others the very means of discovering something 'within himself' which, as he saw it, did not come from 'outside' and could not, therefore, be a possible illusion. If, however, everything that is learned from others and is therefore an experience from 'outside' can be doubted as a possible illusion, may not the language which one has learned from others also be an illusion and the others from whom one has learned it, too? Descartes' doubt, as one can see, did not go far enough. It came to a halt precisely at the point where it might have jolted the philosopher's axiomatic belief in the absolute independence and autonomy of 'reason' as the seemingly ultimate proof of his existence. Evidently the whole problem changes if one does not stop asking questions at this point.

12 The different scenario suggested here brings these hidden contradictions of the old scenario into the open. If one experiments with the 'tabula rasa' hypothesis of a group without antecedent knowledge learned from others, would it still be possible to maintain that, starting from scratch, they could at once perceive and conceptualize connections of events in the form suggested by terms such as 'rational' or 'prior to all experience', as a gift of their untutored reason, as a universal condition of the human 'mind'? Would not, moreover, a regularity such as that of the principle of identity, which is often presented as an eternal 'law of logic' and perhaps even as an appurtenance of people's individual reason, become recognizable simply as a function of people's efforts to communicate understandable signals to others, if a human group is used as the hypothetical point of departure instead of an isolated individual?

If one tries to imagine the experience of human beings without any learned fund of knowledge, one has to take into account their immense preoccupation with the task of satisfying the pressing elementary needs of the moment. Their human potential for synthesis, their capacity for learning connections between happenings, would be realized almost exclusively in the service of these needs; they, their needs and urges, would largely determine what they did and when they did it or, in other words, their timing. On that account alone their experience of the succession of events would be rather different from that represented by the concept 'time'. By no stretch of the imagination can one assume that they might be able to form at once, with the help of the limited experience of a lifetime, the kind of relatively disinterested and impersonal concepts of very high generality, of which 'time' is one.

Even concepts of lesser generality, such as 'moon', 'star', 'tree' or 'wolf' would be beyond their reach. Examine one of them: consider the range of knowledge tacitly implied in the present use of the concept 'wolf'. It includes the knowledge that a wolf is an animal, a mammal, born by a she-wolf, has been young, matured, grown old and will die in a regular, unalterable sequence. This knowledge is almost automatically mobilized by any participant in that knowledge continuum who perceives a shape which he diagnoses as 'wolf'. It endows a person who has learned the word in that sense with a certainty which by now has become absolute, that a wolf cannot transform itself into a man, and vice versa. Neither this knowledge itself nor its certainty could be available to people who do not stand in the line of succession to the flow and momentum of a long continuum of knowledge and of learning. Consider how long it has taken people to gain certainties of this kind: is it 300 years, or less, since the belief in werewolves died out in Europe? How widespread, how firm, is the certainty today that human werewolves are a fantasy? All one can say with great certainty is that the certainty has become greater. The meaning of a concept such as 'wolf' cannot be understood in isolation. It bears the stamp of the overall level which humans have reached in the society where it is used. It can also serve as an example of the way in which advances in knowledge can affect the concept of 'time'. For the emergent recognition of the invariable continuity of one-after-another connections in the case of organisms such as wolves adds yet another type of time-meter to those used before: an organism, and especially a human organism, has come to be seen as a highly specific continuum of changes with its immanent regularities and its relative autonomy filling the interval between birth and death. One may find it useful to remember the specificity and relative autonomy of the processes to which this age time-meter, so-called 'biological time' refers, in discussions about Einstein's contention that people who return from an imaginary space journey may find, on their return, that their friends on earth have aged while they themselves have not.

Examine also the concept 'moon'. For people whose imagination moves within the boundaries of a very wide store of knowledge, it may be hard to imagine that one would not know after a few days' observation the sameness of the thing which appeared in the sky, sometimes in the form of a sickle or a boat and sometimes in the form of a round, yellow face. Would it be possible to imagine that an ancestral group, lifted suddenly from a pre-human to a human condition without any antecedent knowledge, would be able to form at once the idea that the thin, narrow-shaped light which they saw in the sky some time ago and the huge, round face which they see here and now are one and the same thing? Could they, even with regard to such a limited object, automatically achieve a synthesis and produce at once a unified signal – an integrating concept – such as 'moon', for these differently shaped lights in the sky? When and why did it become relevant for humans merely to ask whether the other night's and tonight's light signals in the sky represented one and the same thing? Today the concept 'moon' can easily appear as obvious; one may imagine that one could form it for oneself simply by opening one's eyes and looking at the light in the night sky. The scenario of the human group without antecedent knowledge may make it easier to understand that even so simple an integration of a diversity of sense-experiences into a unitary concept as that represented by the concept 'moon' must have taken a long line of generations to achieve. It could only have resulted from a long process of learning, of the growth of people's stock of experiences, some of which recurred again and again and, over the generations, were remembered as recurring. Apart from the small nexus of events of relevance to the satisfaction of their immediate needs, all that members of an initially conceptless group might have experienced was a welter of vague wandering signals, a kaleidoscopic change, a bewildering coming and going of lights, shapes and other sense-perceptions for which integrating devices, communicable symbolic representations and, thus, means of orientation, were lacking. In their experience, shapes would flow much more easily into each other in an almost dream-like manner. Their ability to distinguish between what we think of as fantasy-experience and reality-experience would be extremely limited. They could hardly know, or learn to know in one lifetime, that there is such a difference as that between dream and non-dream, and what this difference is. Hence the sense of their own identity would be extremely diffuse, compared with ours, and might change greatly during their lifetime.

To those brought up in our type of society, it appears as almost self-evident that every human being has an image of his or her own identity as a being which has been a small child, has grown to maturity, may live to get old and will die. This image of one's own identity as a continuum of changes, a growing and decaying individual human being, embodies an immense range of knowledge; it reflects the relatively high degree of certainty and adequacy which the present knowledge of biological and other regularities has reached today. Without it, one could not have the certainty that one is, as an adult, the same person who had once been a little child. In fact, the conceptualization of processes of change of this or any other kind, as such, is one of the most difficult achievements of people – as one might see from the difficulties encountered today in conceptualizing such long-term processes of change as that of societies or of concept-formation. There is a great deal of evidence to show that the self-image of a person, the sense of his own identity, is more fluid and less firmly organized at earlier stages in the development of mankind. People may experience themselves, and may be experienced by others, as different persons with different names after an initiation rite, or after the assumption of a new social position; they may perceive themselves and each other as identical with their fathers, as transformable into animals or as capable of being in two places at the same time.

Without a long antecedent development of knowledge, people can hardly become immediately aware of the uniformity and regularity of connections over the very wide range known today. Not knowing these regularities, how could they form at once concepts at a very high level of generality, such as 'life', 'nature' or 'reason', now very much taken for granted, but which presuppose a relatively high degree of certainty about the recurrent regularities and the identity of 'objects' as well as of themselves? Is it possible that a group of people who are not heirs to a long antecedent knowledge tradition could ascertain on their own that the biggest lights in the sky (which we now call 'sun' and 'moon') disappear and reappear at more or less regular intervals? Could they conceive, on the basis of the limited experience of a lifetime, the idea that there are such regularities? Again, could they work out for themselves, at the drop of a hat so to speak, classifications such as those of non-living and living things, or of stones, plants, animals and men? How would they work out in one generation communicable symbols of the distinguishing properties of such classifications? Yet, without a reasonably secure knowledge of the distinguishing regularities and continuities of connections between these and other groups of events, would it be possible to conceive the idea of using one continuous sequence of events, such as that of movements in the sky, as a continous timing device for their own social activities?

Consider once more people's sense of their own identity and continuity throughout life. In our type of society the trajectory of a person's life is meticulously timed. A precise social age time-scale – 'I am twelve, she is ten' – is learned very early by every person individually and integrated as a vital element into the image of the individual and others. This numerical time-indexing, however, does not serve as a communication about differentiating quantities alone; it receives its full significance only as a communicable shorthand symbol of known biological, psychological and social differences and changes of people. In the course of a long knowledge process, moreover, the biological and social *processes* to which this age time-scale refers have come to be widely known as unidirectional and irreversible. Hence the *age time-scale* itself often seems to possess the compelling force of an irreversible process; one may speak of 'the passing years' or say that 'time is passing' while one is referring, in fact, to the unidirectional process of one's own life. As is often the case in our type of socio-symbolic universe, highly abstract symbols become reified in common parlance and assume a life of their own. Time-concepts in general, and especially age time-concepts, are particularly prone to this hypostatic use. The continuous series of figures which mark the number of years of a person's life become impregnated with biological, social and personal significance, and thus plays an essential part in a person's sense of his own identity and continuity in, as we say, 'the course of time'.

Moreover, the sense of the irreversible character of the numerical sequence of age-symbols ('twenty-five', 'sixty-four') is heightened by its close connection with the symbols of the era time-scale; that too refers to a unidirectional and irreversible process – to the torch-race of the generations. In relation to the social change-continuum of an era and the corresponding era time-scale, such as that beginning at the assumed date of the creation or of the birth of an era's eponymous hero symbolized, for instance, by the letters BC and AD, the change-continuum of an individual life and the scale used for timing it are exceedingly limited and short. The social process timed by an era scale such as ours (1989, 1999, 2009, 2019) appears to go on for ever. In fact it only goes on as long as the continuity of the corresponding social process is preserved and remembered. The individual processes – people's lives – which concatenate in the form of a social process and are timed by an age time-scale, are sooner or later cut short. The contrast between the shortness of an individual life-span of less than a hundred years and the length of a social era spanning thousands of years, or the majestic length of biological and cosmological processes gradually unravelled by the advancing sciences, reinforces the personal experience which we communicate to each other by means of such expressions as 'the passing years' or 'time wasted and lost'.

Compare the self-experience within such a setting with that of people in a setting where they are not heirs to a knowledge continuum which includes era time-scales, age time-scales or, for that matter, any special devices for timing events at all. Under these conditions one could hardly expect them to experience themselves and their fellows in the way familiar to us. Is it possible to imagine, for instance, that members of the fictitious tabula rasa group coming, as it were, out of the wood, could have at once set about working out age time-scales for themselves, or any time-scales at all? What would give them the idea of making such a time-scale? Yet without it and without a fairly wide and certain knowledge of the regularities of the heavenly bodies as well as their own bodies, would not their self-experience, like their experience generally, differ from ours in a clearly definable and explainable manner? The image people have of themselves, their self-experience in other words, is not, as it often appears today, an item independent of the main body of knowledge, standing apart from their experience of the world at large. It forms an integral part of people's socio-symbolic universe and changes together with it. People's self-image has its place within the trajectory of knowledge leading from the hypothetical condition of absolute ignorance about the connections of happenings tempered by fantasy images of such connections, in the direction towards the lessening of this ignorance and the growth of the reality-congruence of their symbols.

13 'Time' refers to certain aspects of the continuous flow of events in the midst of which people live and of which they themselves form part. One may call them the 'when'-aspects (although that does not cover the whole ground). If everything were at a standstill, one could not speak of 'time'. It is perhaps a little more difficult to see that one could not speak of 'time' in a universe which consisted of one single sequence of changes. If one lived in a single-strand universe of this kind one would never be able to know or even ask *when* anything happened. For 'when' questions aim at pin-pointing events within a continuous flow of events, at fixing milestones indicating relative beginnings and endings within the flow, at marking off one stretch from another or at comparing them regarding their length by means of what we call their 'duration', and other related tasks. All these are timing operations. They could not be undertaken in a single-strand universe. Everything there happens one-after-another; two different stretches of a continuous sequence of changes are never present together. What one does by

means of time-scales is precisely this: one builds milestones and, thus, relative beginnings and endings, into such a sequence with the aid of another sequence. One says, for example: 'we begin at eight o'clock, we end at ten o'clock'; or, comparing durations of stretches within the continuous flow which are not present together: 'we work together for two hours'. It has become possible to do this only because people have found and, later, have worked out for themselves other strands of continuous changes which can serve as standards for timing themselves and the societies they form with each other. At first, continuous sequences of what we call 'natural' events and then, increasingly, human-made mechanical sequences of continuous changes, have served them as means for timing themselves in their three-fold capacity – as biological, social and personal processes. As long as people had not learned to use other sequences of changes as a frame of reference for fixing relative beginnings and endings and, thus, for determining stretches of equal duration within the continuum of changes which they themselves formed, they did not – in fact, they could not – know, for instance, how old they were; they had no common yardstick for the beginning of any recurrent social activity other than that of their own immediate impulses.

Timing thus is based on people's capacity for connecting with each other two or more different sequences of continuous changes, one of which serves as a timing standard for the other (or others). It is an act of intellectual synthesis which is far from simple. For, in substance, the reference sequence can be very different from that for which it is used as a timing standard. The continuously changing configurations of the heavenly bodies are in kind very different from the changing figurations people form with each other. Yet, in one way or another, people have for thousands of years used the former as a means of timing the latter. What specific features have they in common that make it possible to use the one as standard for the other? Or take our watches – in substance a watch is a constantly moving piece of machinery which produces a continuous sequence of changes in the configurations of symbols on its face. As such it bears little resemblance to the continuous sequence of changes in the social and personal lives of people who use watches as a yardstick for articulating their own activities. What relationship does the sequence of changes in the form of watches bear to the continuous social and personal changes of individuals? Different in kind, what have they in common? The answer is deceptively simple – the very fact that they are continuously changing in a fairly regular sequential order (if they stop changing for good, watches cease to be watches and people cease to be people). The concept of time refers to properties which sequences of continuous changes have in common, regardless of their differences in kind.

What we call 'time' is therefore, to begin with, a frame of reference used by people of a particular group, and finally by humankind, to set up milestones recognized by the group within a continuous sequence of changes, or to compare one phase in such a sequence with phases of another, and in a variety of other ways. This is the reason why the concept of time is applicable to quite different kinds of continua of changes. The apparent revolution of the sun about the earth can be standardized in the form of a solar year, which can then be used as a standard of measurement for other cosmic sequences, for a human lifespan or for processes of state formation. This can perhaps be seen more clearly by replacing the substantival concept 'time' by the verbal 'timing' to denote the human activity of synchronization. It is possible to fix positions or segments in the sequence of a horse-race, a chemical reaction, a visit or a war. Sequences at all levels of the universe can be synchronized: at the physical, biological, social and personal levels. This is what is meant by saying that the concept of time is applicable to one-after-another sequences of all kinds, regardless of their specificity. But in all cases of social standardization a particular sequence is needed as a standard, no matter whether it is physical or historical–social in kind. That an investigation of timing makes the sharp conceptual distinction between 'nature' and 'history' untenable is no doubt bound up with the instrumental character of timing; but it also shows that the sharpness of the distinction is epoch-specific and subject to revision.

14 What makes work on the problems of time particularly difficult is that one still very widely attributes to 'time' itself the properties of the processes whose changing aspects this concept symbolically represents. We say 'time is passing' when we refer to the continous changes of our lives or those of the societies in which we live. This peculiar fetish-character of the concept of time is connected to the fact that it represents an intellectual synthesis, a connection of events, at a relatively high level of universality. Not only does it imply the connection, established by men of a particular reference-group, between change-continua within which one attempts to define positions or segments, and a change continuum used as a standard for them, but it also implies the connection between what happens 'earlier' and 'later' in one and the same sequence of events.

In fact, one of the principal clues to the problem of 'time' and of 'timing' is to be found in the specific capacity of people for envisaging together and, thus, for connecting to each other what happens 'earlier' and what happens 'later', what 'before' and what 'now' in a sequence of events. Memory plays a fundamental part in this capacity for visualizing together what does not happen together in actual fact. If I refer to people's capacity for synthesis, I mean especially their capacity for having present in their imagination what is not present here and now and being able to connect it with what happens here and now. This is certainly only *one* manifestation of people's capacity for synthesis, but it plays a crucial part in all forms of timing. To put it in a nutshell: it is meaningless to say that it is four o'clock now unless one envisages at the same time that it was two o'clock before and will be six o'clock later. The very concepts 'earlier' and 'later' are manifestations of people's faculty of visualizing together that which does not happen together and is experienced as such by people.

Here one approaches aspects of the problem of time which reveal its complexity somewhat more clearly. At first glance the concept of time may appear to relate to something monolithic and relatively simple. It may give the impression that the state of affairs it refers to could be defined in a few sentences. In fact, as one can see, it represents a far from simple instrumental connection of events performed by more or less tightly organized groups of men for specific purposes between and within observable continua of change, with or without the inclusion of the continuum which they themselves represent.

So far I have spoken of the synthesis which people perform when they represent a sequence of events as such – when they perceive such a sequence simply as a continuous stream of events which happen one after another. One might call this the integration of sequences in terms of their intrinsic structure. People's own contribution as those who, by their capacity for seeing-together, are able to visualize simultaneously what happens successively and to represent it as a sequence by social symbols, does not enter into this kind of representation. The sequence appears in this case simply as a continuous flow of events which take place, as we say, 'in the course of time'. The symbolization relates to the specific structure of the sequence. People's experience and their synthesizing capacity are not included in this form of the conceptualization of time.

However, there is another set of time-concepts frequently used by us the meaning of which takes into account people's capacity for synthesis, for envisaging together what they know has happened and will happen at different times. I refer to concepts such as 'past', 'present' and 'future'. Their function and meaning are not well understood because one does not clearly distinguish between, and relate to each other, time-concepts such as 'year', 'month' or 'hour' which refer to the temporal structure of the sequential flow as such, and others which include into the orbit of their meaning people as focusing units who visualize the sequential flow and its temporal structure. Past, present and future belong to the later type of concepts. The puzzle they offer to our intelligence is primarily due to their shifting character in relation to the temporal structure of the sequential flow. Today's future is tomorrow's present and today's present tomorrow's past. The solution to the puzzle is simple enough if one gives its due to the specific type of connection which one encounters in any examination of people's mode of experience and to the categorial apparatus that is needed to represent it symbolically. Past, present and future are indicative of the type of concept which is needed as representation of this type of connection. In relation to the sequence of changes which one can represent through the straight series of figures of an era time-scale (1605, 1606, 1607, etc.), the meaning of past, present and future constantly changes, because the human beings to whom they refer and whose experience they represent also constantly change and the connection with them, their experience, is included in the meaning of these terms. What is past, present and future depends on the living generations of the moment. In the intergenerational torch-race, they are always on the move; so, therefore, is the meaning of past, present and future. They too, like the simpler serial time-concepts, such as years or months, are representative of men's capacity for synthesis – in this particular case for experiencing at the same time what has not happened at the same time. But concepts like year, month or hour, though they presuppose this capacity, do not include it in their meaning. They represent only continuous sequences of various lengths as such. The concepts past, present and future include in their meaning the relationship of an experiencing person (or persons) to a sequence of changes. It is in relation to someone who experiences it that one moment of a continuous flow assumes the character of a present *vis-à-vis* others with that of a past or a future. As symbols of experienced time-units, these three terms represent not only – like 'year' or 'cause and effect' – a succession, but also the simultaneous presence of the three time-units in people's experience. One might say that past, present and future, although three different words, form a single concept.

For a long time people have been puzzled by the fact that the actual events within a sequence, and thus the sequential time-units – hours, months, or the years of an era time-scale – to which the term 'present' applies, are constantly changing and that, therefore, the demarcation lines between past, present and future are constantly changing too. The apparent paradox of the three experiential time-units has been noted already in antiquity. Thus Censorinus', starting from a description of 'absolute time', went on to that of past, present and future in the following manner:

[absolute time] is immense, without origin, without end, has always existed in the same manner, will exist in all future and does not appertain to any one person more than to any other. It is divided into three types of time – past, present and future; of these, the past is without entrance, the future without exit, while the intermediary present is so short and incomprehensible that it seems to be nothing more than the conjunction of past and future; it is so unstable that it is never the same and whatever it runs through is cut away from the future and carried over to the past.

Censorinus' formulation of the puzzle, because it is slightly bizarre, can make it easier to recognize the reason for the difficulties. The three terms past, present and future were treated by him almost as if they stood for three different objects 'in time and space'. He approached them, as they are on occasion still approached today, with the categorial apparatus representing connections at the physical level. It is hardly suited to the exploration of connections at the experiential level.

The peculiarity of those concepts of time which include the experience of the synthesis-forming people (concepts such as 'past', 'present' and 'future') is seen perhaps still more clearly if they are contrasted to concepts such as 'earlier' or 'later'. Both can refer to the same sequences. But unlike the kind of timing to which Einstein, for example, devoted his attention, the timing of events as 'earlier' or 'later' is independent of any specific reference group. What happened earlier will always remain earlier than what happened later relatively to it. The concept of the present, by contrast, represents the timing of a living human group sufficiently developed to relate a continuous series of events, whether natural, social or personal, to the change to which it is itself subjected. In accordance with this change, such a group or each of its individual members can distinguish as the present what they are doing here and now, what they are directly experiencing and feeling, both from what is over and subsists only in memory, and from what they may possibly once do, experience or suffer, that is, from the past and the future. It is one of the peculiarities of these time-concepts relating to changes in people living at a particular time that none of them has a clear meaning unless all are present together in the consciousness of people.

Compare terms such as 'earlier' or 'later' used with reference to unidirectional physical sequences – recurrent or non-recurrent – with terms such as 'now', 'today' or 'past, present and future' used with reference to the same sequences. The first kind of concepts, such as 'earlier' or 'later', represents a connection between different positions within a sequence which is the same for all possible reference persons[8], whereas the positions within a flow represented as 'now' or 'present' change if the reference groups or persons change. The demarcation lines between past, present and future are continuously changing because those who experience events in these terms are changing, as I have said before. They change personally on the way from birth to death and, as societies, with the coming and going of the generations (and in many other ways). It is always those living at a given moment in relation to whom events have the character of a present, a past or a future. The social sequence of the 'wandering years' (from one to 2,000 and beyond), like the main sequence of a stellar or a biological evolution, moves along without such landmarks related to particular persons. But there is a difference between social sequences and extra-human, natural sequences in this respect. So far as human societies are concerned, the experience of their process-structure can, *per se*, affect the course of the process. Hence people's experience of social sequences forms part and parcel of the flow of such a sequence itself. With regard to what we call 'nature', that is, the physical level of the universe, that is not the case.

The clarification of the often confusing relationship between time-concepts of the type 'year', 'month' or 'hour' (or 'earlier' and 'later') and those of the type 'present', 'past' and 'future' thus leads to a conclusion which may be unexpected. The latter type of concept does not apply to what we call 'nature' at the physical levels, where the representative kind of connection is taken, rightly or wrongly, to be that of mechanical causality, or only in so far as people relate events at these levels to themselves. The terms 'present' and 'future' can only be applied to the *perpetuum mobile* of the causal chains of 'nature' by virtue of an anthropomorphic identification, that is, figuratively, as when one speaks of the future of the sun. Strictly speaking, this 'nature' is a continuous sequence of changes in the configurations of energy-matter. Within it the helium-pile of the sun only emerges in human consciousness as a phase in a sequence, or as an entity at all, because this formation is of especial importance to human beings. Outside such an experiential relationship the division of 'natural' continua of changes into past, present and future is meaningless.

15 It may help in understanding some of the problems to be discussed later to consider briefly the wider significance of the fact that we work with different types of time-concepts. In the present case there are, on one hand, concepts relating to sequences known by people but not characterized, in their formation, as experienced and known by particular people; on the other, there are concepts which include the experience of such sequences by people (who may be a part of them) in their formation. It is not easy to find appropriate terms to distinguish these two kinds of time-concepts. Perhaps one might speak of 'structural' and 'experiential' concepts. Both are symbols of learned connections or syntheses. But the kinds of synthesis represented by them are different. Terms like 'earlier' and 'later' stand for sequence-related syntheses of positions within a change-continuum as such. They can be applied to mechanical cause and effect connections. 'Past', 'present' and 'future', however, while referring to earlier and later events, are conceptual symbols of a non-causal form of connections – symbols which include in the conceptual synthesis a specific way of *experiencing* sequences. The present, as I have said, is what can be immediately

experienced, the past what can be remembered, the future the unknown that might perhaps come about. Consider the dates AD 1500 and AD 3000. They represent past and future. Between them are years of which people say 'now' or 'the present'. But they take on the character of the present in conjunction with the experience of a past and a future. In the flow of events there are no segments of this kind. What is past merges seamlessly with the present, as the present does with the future. This is seen clearly enough when the future, having become the present, becomes in its turn the past. Only in human experience do we find the momentous demarcations between what is 'today', 'yesterday' or 'tomorrow'.

As examples of time-concepts which articulate the experience of the flow of experience in terms of its relation to the change-continuum of living and experiencing human groups, terms like 'now' or 'present, past and future' are characteristic of the fifth dimension of the universe. With the emergence of human beings the universe takes on, in addition to the four dimensions of space and time, a fifth dimension of consciousness, experience, or however one may express it. Everything which takes place within the purview of human beings now becomes capable of being experienced and represented by human-made symbols, needs in this sense to be defined no longer by four co-ordinates but by five. The developing synthesis which gave rise to the present concept of time is also characteristic of the peculiarity of this dimension. Scientific scruple forbids us to obscure either its natural origin or its irreducible singularity. Yet the epistemological status of experiential concepts such as 'past', 'present' and 'future' which represent this fifth, specifically human dimension, has so far remained uncertain. Discussion of them tends to be regarded as metaphysical; there is a tendency to believe that they cannot and need not be tested against evidence and can be interpreted freely according to personal whim.

That is a misunderstanding. The reason for it is simple enough. A powerful tradition espouses the belief that the categorical apparatus which has been developed in the exploration of the physical level of the universe, and types of synthesis reflecting the forms of connection found on this level, above all mechanical cause and effect connections, are adapted to exploring all the integration-levels of the entire universe, or that there are no other 'rational' types of synthesis. But that is not the case. The power of tradition still makes it difficult for many people today to detach themselves from the notion that processes on the simpler pre-human levels of integration can be the object of precise physical investigations while processes on the far more complex, human levels of integration are inaccessible to the secure methods of scientific study and so are exposed to the caprice of metaphysical speculation. The problems of time have no doubt frequently given rise to such speculations. Yet the developing synthesis represented probably since the origin of states by a concept of time, and the attendant evolution of techniques and institutions of timing, admits very precise analysis and leaves very little room for speculation. 'Time' or, more correctly, timing proves to be a means of orientation elaborated by human beings in the course of centuries in order to perform precisely specifiable social tasks, including the measurement of the movements of heavenly bodies.

It is often difficult to grasp today that the scepticism regarding the possibility of an exact and secure exploration of what has been designated here the fifth dimension, is an over-generalization abstracted from a specific stage – the earliest – in the development of the modern natural sciences. In it one can discover the roots of the idea that every type of connection between events other than those by which classical physics attained its ascendancy, are to be considered meta-physical and unscientific. It is a notion which reflects a particular stage of the knowledge process, and so the status-hierarchy of the sciences or, in other words, the distribution of power between various groups of scientific specialists prevailing at that stage. That studies of the experiential, human–social level tend to be given over to speculation is a hangover from a rationalist tradition which condemns everything not representable in terms of mechanical causal connections to the limbo of irrationality.

What has been said here about the relationship and the difference between the structural and experiential time-concepts can serve as a useful introduction to a change of view. At present the outlook of people is strongly coloured by the unevenness of their knowledge of the world in which they live. They have succeeded in acquiring a great deal of reliable knowledge about the physical levels of the universe. The reality-congruence of the symbolic representations which they use in the exploration of these levels has increased. But their symbols have not got anywhere near to a similar degree of congruence or certainty in their exploration of other levels of the universe, especially of the levels which they themselves represent, the human levels. The adequacy – the reality-congruence – of the symbolic representations which they use in their attempts to explore these human levels and to cope with the problems confronting them there is still very much lower than the fitness of those they have learned to use at the level of 'nature'. Correspondingly limited and uncertain, therefore, is the knowledge they use to orient themselves on these levels.

In fact, the great advances made by people in their exploration of the physical and some biological levels, while they have helped, have also hindered the advance of people's orientation at their own – the human – levels. No doubt the elucidation of these levels has presented specific difficulties of its own. People are obviously afraid of discovering themselves – afraid of what they might uncover. The problems of 'time', too, are often treated like a secret, protective cloak in which to hide. But what comes to light is not the Gorgon's head, nothing terrifying. It is, as will gradually emerge, a picture of the very arduous path that people have covered in attaining a better orientation in their world. They succeeded in replacing the comparatively irregular movements of sun, moon and other stars as a measure of dating, a means of synchronization, by an increasingly close and regular mesh of human-made timing devices; and then, as will be seen, they themselves became so attuned to the synchronization of their behaviour with clocks and calendars that they have come to experience their consciousness of time as a mysterious component of their own nature, or perhaps reified time as a gift of the gods.

16 One of the predicaments of the human sciences, as has just been indicated, is that they are widely dominated by people's attempts to run away from themselves. But these and other inherent difficulties are greatly reinforced by the fact that successful advances in the exploration of 'nature' have given the physical sciences the social status of overall models for all sciences. Those engaged in the study of the human levels of the universe thus find themselves caught in a trap; they are left with two equally sterile alternatives: either they can accept the supremacy of the models of exploration set by the physical sciences without further enquiry into their adequacy for their own task – hoping perhaps in that way to secure for themselves the status of professional scientists; or, finding the models of the natural sciences not entirely adequate for their own task, they can try to work out symbolic representations better suited to the specificity of the connections at the human levels and, in that case, they are apt to founder in a sea of uncertainty – the results of their findings tend to give off a faint odour of speculation and metaphysics.

If one studies today the problems of time, one constantly comes up against the jaws of this trap. As a determinant of causal connections 'time' appears as a clear and precise concept firmly grounded in the physical sciences. The type of connection which one encounters when studying experienced time – *vide* Bergson and Heidegger – is given over to metaphysics. But this is not the only way in which the present differentials in the development of knowledge at the physical and the human levels affect the problems of time. One of the most astonishing aspects of the approaches to the problems of time in our age is the difference between the attention paid to these problems by natural scientists and by social scientists. In the practice of human societies problems of timing have come to play an increasingly significant part: in theories of societies the attention paid to timing problems is, by comparison, minute. To some extent this is undoubtedly due to the fact that weight of opinion has come to attribute enquiries into time to the realm of theoretical physics. If one asked what was, in our day, the most advanced and the most prominent scientific theory of time, one would probably find a great deal of consensus when pointing to that of Einstein. Understandably, sociologists may feel that the problems of time are beyond their reach.

That is not all. The impact of the differential development of natural and of human sciences on contemporary ways of thinking goes much deeper. It finds expression in numerous conceptual dichotomies which make it appear that the world of nature and the world of human beings are two separately and independently existing worlds which in some sense are antagonistic to, or incompatible with, each other. We operate today within an intellectual framework which revolves around conceptual divisions such as 'nature and society', 'nature and culture', 'object and subject', 'matter and mind' or 'physical and experienced time', etc. Although some of them have been handed on to us from previous ages, in their presently dominant version they strongly reflect the developmental differentials of our knowledge and the division of the quest for knowledge into seemingly unrelated academic specialisms. Thus 'nature' has become identified with the field of exploration of the natural, and particularly the physical, sciences. Accordingly, humanity in its various manifestations – society, culture, experience or whatever – has been cast out as that which is not 'nature'. Some may cherish the fond hope that one day all things human may be explainable in terms of the physical sciences. Others firmly believe that an unbridgeable existential chasm opens between humans and 'nature' (as understood by natural scientists). Thus all these dichotomies represent hidden battlefronts. They give the appearance of existential divisions to what reveal themselves on closer inspection to be different values attached by different groups to different, though related, levels of the universe and, by proxy, to the different groups engaged in their exploration. That is the reason why one rests content, in all these cases, with a mere juxtaposition of symbols for different realms of the world in the form of simple conceptual dichotomies, without asking the obvious question – how are these different realms related to each other? We have got into the habit of conceptually splitting the world largely in accordance with the divisions between different academic specialisms.

The dichotomy 'nature and society' is only one of many examples of this kind, but it shows very clearly the deficiencies of this type of conceptualization. By making the two fields appear not only as existentially different, but also as faintly antagonistic to and incompatible with each other, this type of conceptualization quite effectively closes the door to enquiries into the problem of the relationship between what we call 'nature' and 'society'. These two realms are made to appear as independent of each other as the professional academic groups engaged in their exploration wish to be. In actual fact humanity and, therefore, 'society', 'culture', etc. is no less 'natural' – forms no less part of the one universe – than atoms or molecules. 'People' and 'nature' are not really divided existentially in the manner suggested by our present mode of discourse. It is the sciences whose objects of research are 'nature' and 'men' which are divided. Each representative of a scientific specialism tends to perceive his own field of study as an object in isolation; he is apt to attribute to it an absolute autonomy in relation to the objects of other sciences. Thus when scientific studies at the least complex levels, at the levels of inanimate things, gained the ascendancy over all others and proved their worth by the far-reaching applicability of their results to the practical problems of humankind, their field of enquiry under names such as 'nature' or 'objects' came to be perceived as if it existed on its own, wholly divorced from the world of human beings, from the 'subjects' of knowledge. Differentials of the developmental advance of peoples' means of orientation, in other words, projected themselves into the general mode of discourse; they found expression in the different cognitive status attached to 'physical nature' as the structured object of a highly advanced group of sciences and 'people' as a far less structured object of a less advanced group of sciences, and in the seemingly unbridgeable divide separating the 'natural' and the 'human' levels of the universe.

One may like to remember this diagnosis when one speaks of 'two cultures', that of the natural sciences and their application on the one hand and, on the other, that of the social or human sciences and the whole world of art and literature, in short, of 'culture' in the narrower sense of the word. It is quite true that the mode of discourse of the human groups of specialists devoted to these two types of pursuit is very different. Consequently they have difficulty in communicating with each other and, in some cases – let us face it – they are engaged in power struggles, with representatives of each group strongly arguing in favour of the higher value of their own specialism compared with that of another. However, the discussion as it now stands usually gives the impression that both sides consider this split between two groups of specialists speaking different languages and pursuing different goals, as if it were an eternal divide in the world itself, related to the conceptual divisions very familiar to us between subjects and objects, people and nature, history and science. In actual fact it is a very temporary division. As I have already said, it is characteristic of a specific stage in the development of societies at which people have learned a great deal about the handling of inanimate events while their knowledge of themselves as persons and societies, their orientation in their own world and therefore also their understanding of the effects which their more advanced knowledge of inanimate nature has on their social life, lags far behind. The social divide to which we refer when we speak of two cultures is a very temporary condition. To some extent this essay may help to show, and to contribute to, its transience.

It is necessary to indicate here – in broad outline – the insufficiency of the dichotomy 'nature and society' and others of its type because enquiries into the problem of time are effectively blocked as long as they are conducted within the framework of this conceptual dichotomy. It compels the enquirer to treat what one might call 'social time' and 'physical time' – time in society and time in nature – as if they existed and could be explored independently of each other; and this cannot be done. From the first moves human beings made towards timing events, they always were, and always acted as, people *within* and a part of the natural universe. In fact, the problems of time cannot be compartmentalized in accordance with the prevailing compartmentalization of scientific specialisms and the corresponding compartmentalization of our conceptual equipment. Their exploration may thus help to restore the sense of interdependence of 'nature' and 'society' and, in a wider sense, of the unity of the multitudes making up the universe.

It is not particularly difficult to remove the blockage which the conceptual polarity 'nature and society' imposes on our thinking when one begins to explore the problems of time. A good deal of what has to be said later will show this. For the time being, it may be enough to think back once more to the paradigmatic episode which has been quoted earlier – that of the priest who tried to discover for his people the 'right time' for sowing, by observing the passage of the moon through a particular spot on the horizon. Here were people, as they are everywhere, dependent on the fruits of 'nature' for their food. They were dependent on the rain which made the seeds grow; they observed the movement of the moon – a physical movement – in order to find out when it was good for them to sow – a social activity; and they started observing the moon – a social activity – in order to find the best way of satisfying their hunger – a natural impulse.

17 I keep returning to this episode because, so far, written documents have rarely been used as evidence for these early forms of timing; and this episode is taken from one of them. Hence it shows more vividly and directly the function of this form of timing in its wider social context. In other cases one often has nothing but stones as witnesses; they are silent and can easily be misunderstood. Take Stonehenge, the early stone monument on Salisbury plain in Wiltshire, England. It was built in at least three stages sometime during the second millenium by unknown peoples, some of whom may have been of Mediterranean descent. It was evidently a place of worship, an early Delphi of the far west. It consisted, and partly still

consists, of concentric circles formed by rows of huge, upright stones. If one approaches it from the north-east by a road the outline of which is still visible and moved on, as priests or kings and their followers may once have done, in a straight line along the main axis of the circles towards their centre, one can see, looking south-west, again in a straight line with this axis, the sun rising above the horizon more or less directly behind a kind of altar or sun-stone on the morning of the summer solstice. This, in other words, was a place of worship with a built-in timing device. The arrangement is often interpreted as a sign that the people who built this monument were 'sun-worshippers'. But that is hardly more than a cloak for our ignorance. At that stage people usually worshipped gods. They may or may not have regarded the sun as a manifestation of one of them. But whatever the other functions of this monument, it certainly had that of a timing device – more elaborate, yet basically of the same type as that used by the priest who observed the sun's rising above a particular hill as a signal that sowing could start.

Solstice, one may remember, means literally the point where the sun stands still; has reached the farthest point of his summer journey; now he reverses his course and turns back. Many human groups have regarded this turning of the sun as a signal of special significance for themselves. For a very long time people were not at all as sure as they are today of regularities in the movements of the luminous bodies in the sky. There are many reports of peoples being afraid that the moon would not come back after she had disappeared during an eclipse. Perhaps people needed reassurance, too, that the sun would turn their way again after he had travelled in the other direction for a while. They may have felt the need for making sacrifices to their gods to ensure their blessing, or protection against their curse, that is to ensure a good harvest, a rich produce from their herds or victory over their enemies. But whatever the other functions of this monumental setting, its timing device allows one to see the nature of timing and its developmental trajectory in better perspective. Like many other peoples, the people at Stonehenge tried to determine the moment in the sun's movement – we say: 'the time' – when the direction of his course changed in relation to themselves, because they regarded that change as a signal that their own group should embark on a certain course of action. Their 'timing', in other words, was innocently self-centred; it was, in more general terms, group-centred or sociocentric.

18 There is another feature of early timing which gives grounds for reflection; that is its point-like and discontinuous character.[9] One can find a great wealth of evidence to show that active timing in the early stages of its development was confined to selected points of what we know as a continuous sequence of changes. For the time being, the two examples I have given may serve as illustrations. They can, in addition, draw attention to another equally characteristic feature of early timing. Even at a stage at which people already felt fairly certain of their knowledge that the courses of sun and moon in the sky were quite regular, they were for many centuries unable to calculate in advance their changing positions and appearances, by a kind of blind reckoning. If these luminous beings in the sky disappeared for a while, for example, during an eclipse, or changed their form as the moon does when it becomes fat and then thin again, people were for a long time not at all as certain as we are that they would, after a time which could be clearly determined, return to their earlier form or position. They had to see it in order to believe it. Thus the point-like and discontinuous character of their timing went hand in hand with their need for a face-to-face encounter with whatever they used

as a time indicator. They needed to see sun, moon or stars directly in a particular position if their expectation of an answer to their social 'when' questions was to be fulfilled. Their form of timing at this stage is often, rather inaptly, called 'concrete' – inaptly, for even if one can see the new moon and, speaking to others or to oneself, can say, 'there is the new moon', one uses as a means of orientation and communication a concept; and the concept 'concrete', as opposed to the concept 'abstract', can hardly apply to concepts. It is quite possible and, indeed, necessary to distinguish between concepts at different levels of abstraction and generality. Modes of face-to-face timing which are dependent on the direct perception of a point-event, for example, the sighting of a new moon, are representative of a mode of experience at a lesser level of abstraction and generality. In this respect, too, the changes which timing undergoes when societies develop towards the integration of more and more people and increasing functional differentiation are changes in a specific direction.

In the past one has usually tried to solve the problem of 'time' without reference to the developmental trajectory of timing and its overall direction. As long as that is the case the problem remains insoluble. The puzzle of 'time' cannot be solved without reference to the development of the concept of 'time' and of the various time-units, such as 'year', 'month', 'hour' or 'minute', indicating recurrent standard intervals between an earlier and a later position of a unit of change. However, it is possible that access to a developmental approach is blocked by certain evaluative associations which cling to the term 'development' as a legacy of its use in the eighteenth and nineteenth centuries. Whether one speaks of the development of the social institution of timing or of societies generally, the term 'development' is often equated with the old Enlightenment ideal of progress. It seems to convey the idea that every later stage is of higher value than every earlier stage, or a step towards greater happiness. This wishful image of progress is often not sharply enough distinguished from an approach of developmental sociology guided by factually demonstrable progress or, for that matter, regression, in terms, for example, of differentiation or synthesis. One might think of Darwin's approach to the problem of biological evolution. He was not concerned with the problem as to whether amphibians were better than fish, mammals better than reptiles or humans happier than apes; he was simply concerned with the problem as to how and why different species had become what they were and with explaining why species which appeared later in the evolutionary process had functional advantages over earlier types. The problem of the development of societies in general and, in this context, of 'time' in particular, requires a similar approach. As the dominant modes of timing and the corresponding concept of 'time' have *become* what they are today, one can hope to understand and to explain their present condition only if one is able to discover how and why they have developed in that direction.

19 There are a number of issues which play a central part in this essay. They form critical themes which, like leitmotifs, recur throughout the enquiry in a variety of versions. But they often appear unannounced; one cannot always refer to them explicitly without straying too far from the immediate issue, though they may be relevant to it. Hence it seemed advisable to present at least three of them explicitly.

All these problems arise partly or wholly because certain accepted routines of speaking and thinking block access to the problem of 'time' and to much else besides. An enquiry into 'time', as one may have noticed, is a useful point of departure for the great spring-cleaning that is long overdue. There is always a need for it when an intellectual tradition providing the basic means of orientation within its societies has run its course for several centuries, as ours has from the (so called) Renaissance to the present time, from, say, Descartes to Husserl, Galileo to Einstein or, for that matter, from Thomism to neo-Thomism and from Luther to Barth, Bultmann or Schweizer. In all these cases certain basic tenets have become so deeply ingrained in the common customs of speaking and thinking, that they are no longer experienced as something which can possibly be doubted or changed. They are handed on as self-evident axioms by the establishments of the day to those of the next generation, however broken the line may seem on the surface. The longer the tradition lasts, the more self-evident these axioms appear. To bring them to the surface, to examine whether or not the undiscussed 'furniture of the mind' – that which is taken for granted – is any longer serviceable and, if not, to indicate what might take its place, thus becomes an urgent task when a tradition has run its course for some centuries. But it is not a task which any one person can undertake on his or her own. A person on his or her own can have neither the resources of power needed in order to cleanse a society's means of orientation from long-established and deep-rooted axioms, nor the breadth of knowledge or the longevity required for such a task. An individual may even find it difficult to work out and to stand up for new means of orientation which undermine worn-out traditional axioms. But one can have a go at it.

I shall, therefore, select from the broad band of critical issues which are important here three topics which readers of this essay should bear in mind – even in contexts where they are not explicitly mentioned. However, I shall only mention these issues and comment briefly on them as examples of the transition advocated here from a consistently process-reducing to the consistently processual orientation which is really the crux of the matter.

The first issue which I shall introduce concerns a unique characteristic of any standard or meter to be used for determining positions and durations in what we call 'time'. Time-meters differ in certain respects from space-meters. This difference plays a large part in the special difficulties people have experienced – and still experience – in developing time-concepts and time-meters and in formulating clearly the when-problems which depend on such concepts and meters for an answer. Physicists have not overlooked the existence of such a difference between time-meters and space-meters, but the strong tendency towards process-reduction which dominates their tradition, their ideal goal of dissecting change continua into 'isolated systems', is apt to obscure the simple nature of this difference and, with it, the nature of 'time' itself.

The second issue concerns the problem, already mentioned, of the change from a discontinuous, point-like, face-to-face situation-related mode of timing in the direction towards an increasingly closely knit, continuous time-grid of high generality which impinges on peoples' activities throughout the length and breadth of their lives. The social time-grid with which members of highly industrialized nation-states are familiar is of this kind. It is now slowly diffusing throughout the world and the difficulties inherent in the adoption of such a time-grid in areas where, before, earlier forms of timing were in use, are easily observable. One may think that the empirical and theoretical sociological problems which one encounters here deserve greater attention. Anecdotes indicative of the problem can be collected quite easily if one mentions one's interest in problems of time. I was recently told of a South American lady who, on her return from Germany, mentioned as one of her most amazing experiences that there even station clocks had not only two hands indicating hours and minutes, but also a third indicating seconds.[10] Such references to the latest stage so far in the development of timing may help to sharpen awareness of the overall direction and to prepare the way for a more explicit enquiry into the synthesis underlying all time-experiences and time-institutions.

This much is fairly obvious: timing means connecting or synthesizing events in a specific way. Today the synthesis of the temporal aspects of events usually appears as instantaneous, whether it is regarded as derived from one's own experience or believed to be prior to every human experience. To recognize that it is far from instantaneous, that people in the early stages did not and, indeed, could not know how events cohered 'in time' and that it was a painfully slow and laborious intellectual effort to get as far as we have today along the road towards a wider synthesis over-arching the long sequence of the ancestral syntheses, thus leads straight into the heart of the problem of 'time'. The synthesis of which people are capable today, in other words, is a late stage of a very long process. In its earlier stages, nearer to that of the hypothetical conceptless group, people were only capable of a synthesis which was, by comparison, parochial and narrow. This is why their vista, including their timing, appears as discontinuous compared with ours. The second issue thus touches on a wide-ranging set of problems. Not all of them can be explored here.

The third issue that deserves some preliminary comment concerns the relationship between 'nature' and 'society'. That, at least, is the starting point. In our age it is often treated as a self-evident axiom that 'nature' and 'society' are existentially set apart. The problem of 'time' is split accordingly. It seems that physical time is set apart from social time or, for that matter, from experienced time. The examples which I have given so far from simpler societies show that, in their case, timing almost always implies the linking of what we separate as 'natural' and 'physical' and 'social' or 'human' sequences to each other. An examination of the long-term social development of timing thus makes it easier to recognize that the assumption of an existential divide between 'nature' and 'human beings' which dominates many discussions among the learned in our age, is an unexamined axiom of our period, and one has to examine whether it is not a period-piece.

It is particularly relevant to raise this question in this context for, as long as the axiom of the split world is accepted as self-evident, one cannot come to grips with the problem of the relationship between the 'natural' levels (which, according to present conceptual usage, usually means 'physical' levels, but may also mean 'biological' levels) and the 'human' ('social', 'experiential') levels of the universe. One remains tied to the prevailing notion that 'time' is the concern of physicists, especially of theoretical physicists, while the sociological problem of 'time' is left fallow in the no-man's land between the natural and the social sciences. The tendency of each group of scientists to treat its own domain as sacrosanct, as a fortress protected from alien intruders by a moat of common professional conventions and ideologies, stands in the way of any attempt at linking different scientific domains to each other by means of a common theoretical frame of reference. As things are, it is difficult to dismantle these barriers when one is concerned with the problem of 'time'; it is difficult, in other words, to think and speak in a manner which does not tacitly imply the assumption that physical time, biological time, social and experiential time stand unconnected side by side. That is why it is necessary in this context to examine the peculiar split which, in our tradition, runs through the whole symbolic representation of the universe and of which conceptual polarities such as that between 'nature and society', or 'object and subject' are examples. If one explores 'time' one explores people within nature, not people and nature set apart. That, too, is a recurrent theme of this essay.

20 I shall now examine in somewhat more detail the first of the critical issues I mentioned earlier: the relationship of 'time' and 'space'. In this discussion, however, a wider issue is at stake. An ageing tradition has allowed layer upon layer of intellectual grime to obscure and to erode the core of its intellectual orientation. As one removes them step by step, what is underneath comes to light. This is one such step.

The concepts 'time' and 'space' are among the basic means of orientation of our social tradition. It is easier to understand their relationship if one, once again, relates the substantives to the corresponding practices. 'Time' and 'space' are conceptual symbols of specific types of social activities and institutions; they enable people to orient themselves with reference to positions or intervals between such positions which events of all kinds occupy in relation both to each other within the same sequence of events and to homologous positions within another standardized sequence of events. To perceive and to determine spatial and temporal positions as such, therefore, becomes possible only at a stage of social development at which people have learned to handle and reflect on events with the help of means of orientation at a relatively high level of generalization and synthesis. Positional relations in 'time' and 'space' are those relations of observable events which remain when one abstracts from all other possible relations of events in a specific order of magnitude – in that, say, of galaxies and grains of sand, or of whales, humans and baccillae – and then connects or 'synthesizes' the remainder. The two terms thus stand for purely positional relations of observable events at a very high level of abstraction and synthesis.

That human beings have developed two different concepts for connection at this level is due to the fact that the type of meter required, in practice, for determining relations of temporal positions is in some respects different from that required for determining relations of spatial positions. Positional relations 'in space' (as we call it) are those which can be determined by means of non-moving and unchanging standards, although people, in order to use them as meters, may have to move them about and to change their position 'in space' and thus 'in time'. It is purely in relation to people that certain sets of related positions are non-moving and are thus conceptually separated as 'spatial' from others which, in relation to people, are on the move. Positional relations in what we call 'time' are those which can be determined only by means of moving standards, of meters which are in a condition of continuous positional change. The conceptual separation which makes it appear that 'time' and 'space' are different and perhaps even separate units is thus simply the result of an attempt to distinguish conceptually between two types of purely positional relationships, those which can be determined by non-moving meters and others which can be determined only by moving meters in a condition of continuous positional change. One need only think of the function as 'time-meter' of the continuously moving sun, or of watches and calendars (which represent a continuous positional change symbolically even if they themselves do not move) and compare them with 'space-meters' such as rulers or milestones, in order to see the difference and to take one's first step on the road towards an explanation.

The difference between the two types of meters tells us something of the difference between the practical tasks involved in determining purely positional relations of the spatial and the temporal type. It is, in the first instance, these practical differences which account for the conceptual distinction between positional relations 'in time' and 'in space'. The positional relations themselves are entirely inseparable. The story of the conceptual reunion of 'time' and 'space' culminating in Minkowski's and Einstein's concept of a four-dimensional continuum need not be told here. What needs to be said in this context can be expressed in relatively simple language. In brief: every change in 'space' is a change in 'time'; every change in 'time' a change in 'space'. Do not be misled by the assumption that you can sit still in 'space' while 'time' is passing: it is you who are growing older. Your heart is beating, you are breathing, you are digesting; your cells are growing and decaying. The change may be slow, but you are continuously changing in 'space' and 'time' – on your own, while growing and growing older, as part of your changing society, as inhabitant of the ceaselessly moving earth.

Space-meters can be non-moving or, for that matter, non-changing in relation to people, as they are selectively standardized for determining relationships between positions which also do not change relative to people who measure them. In that way it becomes possible to lay these standardized scales of relationships alongside the positions whose relationship one seeks to determine. To answer where-questions in terms of a chosen yardstick is comparatively simple because the yardstick representing specific positional relationships, as well as the objects to which it is applied, stands still in relation to certain people. To determine relationships between positions of events which happen one after another, which form part of a continuous sequence of change, without reducing them conceptually to relationships between positions which stand still is a much more difficult task. As the universe in which human beings live and of which they form part is continuously moving and changing, one way of accounting for the conceptualization of 'space' and 'time' as different, or even as separate, is to say that what we call 'space' refers to positional relationships of moving events which one tries to determine by abstracting from the fact that they are moving and changing; by contrast, 'time' refers to the relations of positions within a change-continuum which one tries to determine without abstracting from its continuous movement and change.

21 At present one can easily conceal from oneself the difficulties encountered if one does not fall in with the demands of an intellectual tradition based on the selective reduction of changing and moving to unchanging and non-moving relationships. For the pressure to conform to the traditional mode of speaking and thinking in change-reducing terms is exceedingly strong. Even the concept of 'time' can succumb to this pressure. While being used in one sense as an overall symbol for irreversible positional changes from an earlier to a later position, it is used simultaneously, for example in physics, as a name-giving symbol for unchanging quantities on a part with quantities of non-moving aspects of events such as space or weight. Thus the term 'time' is used today with reference partly to a continuous sequence of unrepeatable changes and partly to repeatable sequences of change which, in so far as they are repeatable, are unchanging. Yet neither the difference nor the relationship between these two types of 'time' has been clearly set out or explored as an open problem. In this case, too, the routinized use of the term 'time' conceals the unsolved problem behind it. One can understand better what this problem is and, thus, what we mean by 'time' if, on occasion, one avoids using the term 'time' when custom exerts pressure to use it, and examines the problem which people try to solve with its help.

An example may show the problem in better perspective. Let us assume that a people whose members traditionally assemble in order to take political decisions by vote, after hearing speakers of opposite factions arguing for and against the decisions, have made a rule that opposite speakers must take up their position at a certain distance from each other. Distance, in this case, means a specific relationship between two positions 'in space'. It is relatively easy to determine this type of positional relationship and to check whether the two speakers obey the rule. One can do it simply by standardizing the length of the interval between the two positions. A piece of wood or a rope standardized in this manner and laid out between the two speakers can serve this purpose. But assume the rule is that two antagonists speaking at the assembly must not make speeches beyond a certain length. How does one determine the distance between two positions, that is, between the beginning and the end of speeches which occur one after another? How can one compare the interval between the beginning and the end of something which is continuously moving (a man's speech) with that between the beginning and end of another continuous change (another man's speech) which takes place before or after the other? The two speeches can never run simultaneously without losing their function as interdependent communications. How, then, can one make sure that the one does not run on longer than the other? The ancient Athenians were faced with this problem (which might not have arisen if the solution had not been at hand). They solved it by means of an hourglass which they standardized so that the sand running out indicated clearly when a speaker was running on beyond the appointed limit, longer than the other. We say, in this case, that the Athenians measured the 'time' allowed for each speech. However, if one does not immediately cover up the problem by saying they measured 'time' and insists on finding out what it actually was the Athenians measured, one can see quite clearly the hidden problem behind the word 'time'; for as 'time' is not an occurrence directly accessible to sense-perception, how can one measure it? What does one actually do if one says that one measures 'time'?

Take as example what the Athenians did: they compared the interval between the beginning and end of the flow of two speeches or, in other words, of two non-repeatable social change-continua which occurred one after another, with the interval between the beginning and end of the flow of sand in an hourglass – of a repeatable change-continuum which, according to our scheme of things, is classified as 'physical' or 'natural'. By saying the Athenians 'timed' the speeches, one simply means that they rendered the intervals between their beginnings and their ends comparable by reference to the intervals between the beginnings and ends of inanimate processes which, in contrast to the speeches, were repeatable and more reliable and controllable than sequences of human actions. Here, as one can see, timing was entirely sociocentric: one used repeatable, inanimate sequences of limited length as meters for non-repeatable social sequences. Throughout the earlier stages of the development of timing, the use of intervals between two positions of a physical movement, whether it was the movement of sand in an hourglass or of sun and moon in the skies, was a means to an end. It had an instrumental character. Observations of the changing positions in the sequence of physical events in the sky or on earth were not undertaken for their own sake; they were used as indicators which told people *when* specific social activities ought to be undertaken or *how long* these activities ought to last.

If one says today in this context that the Athenians' hourglass was a means of measuring 'time', one uses the term 'time', to begin with, as a useful symbol for something very tangible, for example, the length of two successive speeches. Here the verifiable and standardized change-continuum of the falling sand served as the socially standardized relational sequence. But the hourglass can, of course, also be used as standard sequence for many other change-continua, such as the running of a certain distance or the boiling of an egg. Whether hourglass, sundial or quartz clock, time-meters are instruments which people have created for quite specific purposes. They serve as common relation-sequences for determining successive positions in a multiplicity of often very different sequences of events, each of which is potentially or actually visible and tangible, like the sequence of the clocks themselves.

The common feature of this multiplicity of specific sequences of events that people seek to measure by means of clocks, or calendars, is called 'time'. But because the concept of 'time' can refer to when-aspects of very different sequences, it is apt to appear to people as if 'time' is something existing independently of any social standardization of relational sequences and of any relation to specific sequences of events. They say: 'We measure time' when they are trying to synchronize, to date – in a word, to 'time' – aspects of quite specific potentially or actually tangible sequences of events.

And this fetish character of what we call 'time' is particularly reinforced in people's perception because the social standardization of individuals in terms of socially institutionalized time is anchored more firmly and deeply in their consciences the more complex and differentiated societies become, and the more necessary it becomes for each individual to ask himself constantly: 'What is the time?' or 'What is the date today?' It would not be difficult to show how, on the way from, say, water, sun and sand clocks to church-tower clocks and from there in the course of centuries to the individualized wrist-watches worn by individuals, the co-ordination of the behaviour and feelings of individuals to socially institutionalized 'time' becomes more complex and self-evident. If social demands were met earlier by a crier or bells summoning the faithful morning, noon and evening to prayer, at a later stage it was necessary for public clocks to show the hours, and at still later stages of social development, the minutes and even the seconds.

The use of clocks to measure purely physical sequences started only shortly before Galileo, and was really only introduced by him. That, in other words, physical time only branched off fairly late from social time, has already been mentioned. However, physicists and philosophers lost sight early on, in their specialized reflections, of the connection between their knowledge and its native soil of developing human society. In addition, the structure of this development was devalued through being misunderstood as structureless 'history', as something 'merely historical'. Sociologists, for their part, concerned themselves little with 'time'. So 'time' remained a mystery.

22 One of the decisive turning-points in the development of timing was the branching off from the older, people-centred mode, of a nature-centred mode of timing; but the changeover did not occur at once – it was a long and slow process. Even Ptolemy, whose work represented the synthesis of all the knowledge of astronomy acquired by the more developed ancient societies of the Middle East and eastern Mediterranean distinguished in no way as sharply as we do today between the immanent regularities and connections of the movements of celestial bodies and their meaning as indicators of people's destiny. He wrote a treatise on astrology as well as on astronomy; and in his own eyes, as well as in those of his contemporaries, they evidently supplemented each other. Enquiries into the positions and movements of celestial bodies were still closely linked to enquiries into the meaning which these movements had for men. 'Nature' and 'humanity', 'objects' and 'subjects', did not yet appear as two existentially divorced realms of the universe.

The common intellectual framework of Medieval scholars, too, lacked the notion of an existential divide of this kind. If they spoke of 'nature' they thought of it as an aspect of God's creation. Within it human beings had an especially high rank and value. But it also included animals, plants, sun, moon and stars. The heavenly bodies, nearer to God, were without blemish; their movements were faultless and circular. Earthly bodies, in movement artificially disturbed by human or beast, were credited with a built-in aim to return to their 'natural' place – the place of rest allotted to them in God's creation. The consistent use of a God-centred frame of reference compelled people to approach the problem of 'nature' and, thus, of physical movements, in the last resort by means of concepts fashioned so as to account for the aims implanted by God in all his creatures, for the ends towards which they were designed by God to strive. The concept of a theocentric universe induced a teleological concept formation. Hence the meaning of the concept 'nature' as used by Aristotle and the Medieval scholars was different from that which gained ascendancy when the God-centred frame of reference and the social institutions which backed it lost their hegemonial power.

The meaning of the concept of 'time' changed in the same way. The study of physical sequences for their own sake and a consistently 'nature-centred' mode of timing came into its own from the time of Galileo on. Although Galileo in no way abandoned the belief in a God-centred universe, he consistently abandoned the concepts of nature which bore its imprint when he studied problems such as that of the trajectory of cannon balls or the velocity of falling bodies. He no longer worked with the concept of a 'natural' place towards which these inanimate units were striving or with the distinction between a 'natural' or God-given and a violent, for example human-induced, motion which also bore the imprint of a God-centred frame of reference. Instead he set out to discover the immanent regularities of observable connections between events – regularities which, inexplicably, could be represented by mathematical equations and, thus represented, soon acquired in people's eyes the status of eternal 'laws' underlying all the observable changes of 'nature'.

Thus, once more, the quest for something permanent, unchanging and eternal behind the continually changing course of observable events asserted itself. As symbol of that which remained always the same no matter how the world might change, 'God' was gradually a new concept of 'time' – 'time' as a quantifiable and indefinitely reproducible invariant of the 'laws of nature' – branched off from the once relatively[11] unitary, human- and God-centred concept. Like 'nature' itself, 'time' became increasingly mathematized. It became one of the concepts which, like weight, spatial distance, force and many others, could be measured in isolated doses quite independently of the 'time of day' (or of the week, month or year). Increasingly, humans engaged in their scientific studies got into the habit of saying they were measuring 'time' without any attempt at investigating the observable data to which this concept referred, even though 'time' as such is hardly visible or tangible, that is neither observable nor measurable. For this reason it can neither expand nor contract.

23 Galileo's work illuminates most vividly the new turn taken by the concept of 'time' in its development from the Middle Ages on, just as it illuminates the emergence of the new concept of 'nature'. It may be useful to read, in Galileo's own words, a description of his famous acceleration experiments which helped to launch the development of the 'nature-centred', the physicists' concept of 'time'. Galileo's description shows better than any paraphrase the immense effort he put into his experiments and the high value he attached to them:

A piece of wooden moulding ... about twelve cubits long, half a cubit wide, and three fingerbreadths thick, was taken; on its edge was cut a channel a little more than one finger in breadth. Having made this groove very straight, as smooth and polished as possible, we rolled along it a hard, smooth and very round bronze ball. Having placed this board in a sloping position by lifting one end some one or two cubits above the other, we rolled the ball, as I was just saying, along the channel, noting, in a manner presently to be described, the time required to make the descent. We repeated this experiment more than once in order to measure the time with an accuracy such that the deviation between two observations never exceeded one-tenth of a pulse beat. Having performed this operation and having assured ourselves of its reliability, we now rolled the ball only one-quarter the length of the channel; and having measured the time of its descent, we found it precisely one-half of the former. Next we tried other distances, comparing the time for the whole length with that for the half, or with that for two-thirds, or three-fourths, or indeed for any fraction; in such experiments, repeated a full hundred times, we always found that the spaces traversed were to each other as the squares of the times, and this was true for all inclinations of the plane, i.e. of the channel along which we rolled the ball . . .

. . . For the measurement of time, we employed a large vessel of water placed in an elevated position; to the bottom of this vessel was soldered a pipe of small diameter giving a thin jet of water, which we collected in a small glass during the time of each descent, on a very accurate balance; the differences and ratios of these weights gave us the differences and ratios of the times, and this with such accuracy that although the operation was repeated many, many times, there was no appreciable discrepancy in the results.[12]

Galileo's experimental strategy was as simple as it was ingenious. It shows very clearly what it was he measured when he said he measured 'time'. He used as a meter for the movement of downward rolling balls a thin jet of downward moving water. He measured, by weighing it, the amount of water that flowed through his water-clock while the ball was moving from its resting position A to a position B and compared this amount with that flowing through his clock while the ball was passing downward from B to C and from C to D. Measured with a ruler, that is in purely spatial terms, the distances AB, BC and CD were of equal length. However, the amount of water running down in the time-meter while the ball moved through AB was, as Galileo found, not the same as that running down while it passed through BC and, again through CD: it became progressively smaller. The observaion that progressively less water flowed through the clock while the ball passed through any given later distance than flowed when it passed through any equal earlier distance was expressed by Galileo in accordance with the social function of a clock, in terms of 'time'. Decreasing quantities of water, representing shorter runs of water through the water-clock, were the observable data on which was based the statement of decreasing 'times' for equal distances of the downward moving ball. The term 'time', in other words, stood for a physical movement – that of the time-meter. Unequal amounts of water running down the pipe of the clock and observed by Galileo on the scales were seen by him as indicators of 'unequal quantities of time'. In that way his experimental strategy gave him the certainty that the ball moved through later intervals of its journey more quickly than through spatially equal earlier intervals or, in other words, that the 'time' of its passage through earlier was longer than that through equal later intervals, that its speed was not uniform: it accelerated.

Galileo could not have discovered with certainty the accelerating speed of a body's fall or, by proxy, a ball's movement down an inclined plane, without a relatively uniform movement as a frame of reference – without a 'time'-meter. As one can see from his description, he still occasionally used in his acceleration experiments one of the simplest time-meters at people's disposal – the pulse-beat. Though not highly reliable, its pattern is normally that of a non-accelerating movement with a succession of strongly marked intervals of equal length. Neither water-clock nor pulse-beat, used by Galileo as common meters for positional changes which happen one after another, had that measure of regularity and uniformity to which we are used today in timing events. But their movement was sufficiently uniform for him to ascertain that the speed of a falling body, the 'time' of its change from position to position, was not uniform; it even enabled him to find out exactly how much the speed of a falling body increased or, in terms of a time-meter, how much the 'time' of its passage through spatially equal intervals decreased in the course of its journey. By means of his experiments, Galileo tested and confirmed his assumption that the distance fallen was proportional to the square of the 'time' or, in the form of an equation:

$$D \sim T^2$$

Neither these experiments nor the regularities they were designed to test were conceived by Galileo in a flash of inspiration. Many years of reflection and observation had led up to them. In the course of this long process of discovery he had followed wrong tracks and strayed into blind alleys. But he persisted; he was flexible enough to recognize his own mistakes and adventurous enough to look for better solutions. In the end he succeeded – to a point.

Later generations have usually interpreted Galileo's work in accordance with their own preferences. He became, as the dead usually do, the prey of his interpreters and, often enough, a crown witness for the defence of their own beliefs. Thus, to empiricists he appeared as an empiricist and to idealists as an idealist – a Platonist. People have quarrelled endlessly as to whether he first conceived what was later called the 'law of falling bodies' deductively, that is purely by reflecting, or inductively, that is purely by observing. Yet very obviously these questions are absurd. They imply that people can reflect without antecedent observations and observe without antecedent reflections. No known procedure of research and discovery in the positive, the theoretical–empirical sciences corresponds to either of these assumptions. The two antagonistic views are cut according to the same pattern. They share the assumption that one has to find a beginning, that discovery *begins* either in reason or in experiment. Both sides, like other antagonists at the same stage of social development, are bound to each other by the same problem to which they give opposite answers. They ask where advances in knowledge begin. They aim at a static explanation modelled on the causal explanations of physics. Yet discovery is a process. In the reconstruction of such a process, in the building up of process models, analysis – the dissection into factors or causes – is an auxiliary to synthesis. There are no beginnings – no reasonings without observations, no observations without reasonings.

What is usually not seen clearly enough is that these are the battlefronts of a later stage. Galileo's own battles were fought on a different front and his own position, therefore had a different construction. The principal controversy which helped to shape his own views was the controversy with those who had transformed Aristotle's teaching into an authoritative and final doctrine which could not be doubted and for whom, in course of time, these views had assumed almost the same character as the holy books of Christianity, the depository of revealed knowledge.[13] According to Aristotle, the velocity of a falling body was proportional to its weight. Galileo's efforts were designed to refute that view and to replace it with a theorem more solidly founded on testable evidence. His criticism was directed not only against the substance of the Aristotelian doctrine, but also against the fact that its followers tried to test Aristotle's theories by means of systematic experiments.

Galileo was certainly not the first to recognize the innovatory potential of reflection bonded to systematic observation and experimentation. Florentine society had been a fertile breeding ground for the 'new realism', for innovatory attempts at breaking away from established tradition through individual reflection and experimentation aided by systematic measuring techniques. Perspectivist painting, developed within the circle of experimenting masters around men like Masaccio and Uccello and aided by the more theoretical reflections of Alberti, is an example. Galileo's work represents a culminating point in that tradition. To characterize it as representative of either 'empiricism' or 'idealism' is an anachronistic misconstruction. The 'new realism' of his age, directed against the Aristotelian orthodoxy, meant developing abstractions in such a way that they became, relative to the antecedent stage, *more* reality-congruent. Present knowledge of the long process of observation and reflection leading up to Galileo's discovery of a more realistic formula for the movement of falling bodies is still fragmentary; but one can see clearly enough that practical, theoretical and empirical problems together contributed to it.

One of the practical problems that interested Galileo was that of the functioning of weapons such as cannons. Knowledge of the regularities of falling bodies helped him to analyse projectile motion and to calculate the trajectory of a cannon ball. By learning more about these regularities, Galileo was better able to make suggestions for the improvement of ordnance. It is not entirely accidental that he used the arsenal of Venice as a locale for discussions about his new science. He proudly declared that, although many people before him had noticed the curved trajectory of missiles, no one before him had known that it had the form of a parabola. As such, it could be determined and, if necessary, manipulated with the help of fairly rigorous mathematical calculations. A parabola, as Galileo saw it, could be regarded as a combination of a forward and a downward movement. Mathematical techniques could thus be applied to practical ballistic problems if one knew the immanent regularities of these movements.

The problem of the regularities of the movement of falling bodies, with which Galileo wrestled for many years, had its place in this context. Was the speed of the fall proportional to the distance? – to the time? – to the weight? Was the speed uniform? Did it change? Did it change uniformly? It was characteristic of Galileo's penetrating ingenuity that he recognized the general theoretical problem inherent in the specific practical problem. The task, as he saw it, was to discover purely and simply the recurrent relationships between different aspects of the movement of falling bodies – different aspects whose quantitative values had to be determined with the help of different meters. He completely abandoned the ultimately teleological frame of reference within which his precursors had discussed the problem of physical movement – largely, one may think, because this frame was irrelevant to the practical problems he had in mind. In Galileo's case, thus, practical and theoretical problems complemented each other; they were not divorced, as is often the case today, nor yet entangled in priority quarrels as to which was the cause and which the effect.

In exploring such problems Galileo, like everyone else, was bound by the limitations imposed on him by the technical and, in a wider sense, the social development of the period in which he lived. These limitations showed themselves clearly enough in the way he tried to determine the speed of free-falling bodies, which was relatively great. Timing devices with subdivisions, such as minutes and seconds, which were small enough to allow accurate measurement of movements at that rate of speed had not yet been invented. It is not unlikely that this was one of the reasons why he tested his ideas about the regularities of free-falling bodies indirectly, with the help of balls rolling along an inclined plane. It slowed down the speed of the downward movement and kept it within the limits compatible with the speed of the movement of the timing devices available to Galileo. This is a vivid reminder of the fact that measurements of the 'time' and the speed of movements are functionally interdependent. Over the same distance, the greater the speed the shorter the 'time'. If one knew the one, one knew the other.

Galileo's empirical problem was how to find reliable empirical evidence for the 'time' distance ratio of downward moving bodies. The experimental strategy outlined before was his answer to this problem. It enabled him to determine the speed or the 'time' of movements over given distances with greater accuracy than any of his predecessors and, thus, to test the old hypotheses and his own new hunches about the movements of falling bodies in a way no one had used before. This, too, the working out of a new experimental strategy, had taken him a long time; it was the result of a lengthy process in which both reasoning and observation played their part. In all likelihood he had made rough and ready experiments of this kind, using perhaps his pulse-beat as a time-meter, before he worked out the ingenious experimental arrangement which he described in the passages that have been quoted.

As far as we know, never before had human-made time-pieces been used in this manner as a measuring rod for physical processes.[14] The clepsydra, an elaborate version of which he used in his experiments, was traditionally a timepiece employed for timing human affairs. It was a social time-meter. Timing had been human-centred. Galileo's innovatory imagination led him to change the function of the ancient timing device by using it systematically as a gauge not for the flux of social but of natural events. In that way a new concept of 'time', that of 'physical time', began to branch off from the older, relatively more unitary human-centred concept. It was the corollary of a corresponding change in people's concept of nature. Increasingly, 'nature' assumed in people's eyes the character of an autonomous, mechanical nexus of events which was purposeless, but well-ordered: it obeyed 'laws'. 'Time' assumed the character of a property of that nexus. A long, slow development had prepared the way for this branching off, from the earlier God- and human-centred manner, of a nature-centred manner of timing. Looking at Galileo's experiments one catches, as it were *in vivo*, the moment of departure.

24 The significance of this emergence of the concept of 'physical time' from the matrix of 'social time' can hardly be overrated. It went hand in hand with the emergence of a new function for human-made timepieces; it implied the timing of 'nature' for its own sake. Hence it was one of the earliest steps in a process of concept-formation whose results today have become fossilized and are very much taken for granted – steps on the road towards the conceptual split of the universe which has come to dominate increasingly people's modes of speaking and thinking and which appears as a consensual axiom that no one can doubt. As an autonomous nexus represented by eternal laws, 'nature' appears to stand on one side, people and their social world – artificial, arbitrary and structureless – on the other. Endowed with regularities of its own, 'nature' as an object of people's studies seems to be, in some way not clearly explained, divorced from the world of humans. One has not yet come to recognize that the illusion arises from the very fact that humans have learned to distance themselves, in their reflection and observation, from 'nature' in order to explore it – to distance themselves more from 'nature' than from themselves. In their imagination, the greater distancing and self-discipline required for the exploration of the inanimate nexus of

events transformed itself into the notion of a really existing distance between themselves, the subjects, and 'nature', the nexus of objects.

The growing dualism of the concept of 'time', the emergence of which one can witness by envisaging Galileo's acceleration experiments in this wider context, reflected most graphically the growing existential dualism of people's image of the world at large. In the societies afflicted by it, this dualism has taken root so firmly that its members have come to expect, as a matter of course, to be able to classify events as either natural or social, objective or subjective, physical or human. In connection with this wider conceptual divide 'time', too, came to be divided into two different types: physical and social 'time'. In the former sense, 'time' appeared as an aspect of 'physical nature', as one of the unchanging variables which physicists measure and which, as such, plays its part in the mathematical equations intended as symbolic representations of nature's 'laws'. In the latter sense, 'time' had the character of a social institution, a regulator of social events, a mode of human experience – and clocks had that of an integral part of a social order which could not work without them.

In developmental terms, this division into physical and social 'time' was closely associated with the rise of the physical sciences. As they gained ascendancy, 'physical time' came to be regarded more and more as the prototype of 'time'. According to a value system, of which more will have to be said later, 'nature', the object of the physical sciences, appeared to people as the epitome of good order and thus, in some sense, as more 'real' than their own, seemingly less orderly and more haphazard social world. Physical and social 'time' were evaluated accordingly. 'Physical time' could be represented in the form of isolated quantities; it could be measured with great precision and quantities of 'time' could take their place together with the results of other measurements in mathematical calculations. Constructing theories of 'time', therefore, came to be regarded almost exclusively as the task of theoretical physicists or of philosophers as their interpreters. In contrast, 'social time', though its significance in the practice of people's social life steadily increased, seemed to have hardly any significance as a theoretical concern or, more generally, as a subject of scientific enquiry. In an almost perverse distortion of the actual sequence of events, it appeared as a somewhat arbitrary derivative of the more firmly structured 'physical time'.

Thus the conceptual dualism went hand in hand with clearly marked differences in the status and value attached to the two types of 'time'. The very term 'natural time', compared with that of 'social time', conveys the idea that the former is 'real', the latter an arbitrary convention. The difficulty is that 'time' does not fit into the conceptual schema of this dualism; like a good many other data, it defies classification as either natural or social, either objective or subjective: it is both rolled into one. One of the principal clues to the persistence of the apparent mystery of 'time' lies in the persistence of this conceptual division. The puzzle cannot be resolved as long as the division between 'nature' and 'society' and, thus, between 'physical time' and 'social time', characteristic of the present stage of development, is understood as an eternal, existential division – and as long, therefore, as the problem of the relationship between 'physical time' and 'social time' remains unexplored.

In the social context 'time' partakes of the curious mode of existence of other social data represented by nouns, such as 'society', 'culture', 'capital', 'money' or 'language', which refer to something that, in some undefined sense, appears to exist outside and apart from human beings. On closer examination one discovers that nouns of this type, including 'time', refer to data which presupposes a plurality of interdependent human beings and, for that reason, have a relative autonomy and may even have a power of compulsion with regard to each of them singly. Individually, therefore, people often have the illusion that social data of this kind, because they are independent of themselves as individuals, are independent of human beings generally. Especially in urban societies they make and use clocks in a manner reminiscent of the making and using of masks in many pre-urban societies: one knows they are made by people but they are experienced as if they represented an extra-human existence. Masks appear as embodiments of spirits. Clocks appear as embodiments of 'time'; the standard phrase used with regard to them is: they indicate 'time'. The question is: what exactly do clocks indicate?

It may help towards a better understanding of the functions of time-meters if one considered once more their distinguishing characteristics. I have already mentioned that they differ from other measuring devices, such as those concerned with length or weight, in that they are continuously moving. But that alone does not do justice to their distinctness. One has to add that their movement is relatively independent of that which they measure and that it is a unilinear (if circular), unidirectional (if repeatable) movement with uniform velocity, that is, a non-accelerated movement. The idea that clocks 'indicate' or 'register' time is open to misunderstanding. *Clocks (and time-meters generally) human-made or not, are simply mechanical movements of a specific type, employed by people for their own ends.* From the use as a timing device of the sun moving through the constellations, to that of the hands of a pendulum clock moving through the numerical symbols on the clock-face and to the 'micro-wave' oscillator of an atomic clock displaying 'time' on an electrically driven 'dial', timing devices always have movements of their own with a pattern that is characteristic of their function. They all proceed at an even speed through a continuous succession of changing positions in such a way that the length of their passage through any earlier interval between two successive positions equals that of their passage through any equidistant later interval. As the duration of their passage through earlier and later intervals is the same, they can be used in a great variety of ways – as meters, as means of orientation, as frames of reference for a multitude of other sequences of change which proceed from an earlier to a later position in an unrepeatable and less regular manner. By means of these and other timing devices one can, as it were, build milestones into the continuous flux of other sequences, where it is not possible to compare with each other the durations of successive changes from one position to another. Seconds, hours, twenty-four-hour days, or any other subdivision of a time-meter's continuous movement, follow each other in a unidirectional line of succession. Within this sequential order, every second, every hour or day, is unique and unrepeatable; it comes and goes and never returns. But the duration of a movement's passage between two positions that is socially standardized as 'second', 'hour', 'day', 'month' or 'year', is exactly the same as the duration of a movement's passage standardized in the same way: one 'second', 'hour', 'day', etc. lasts as long as any other, although it is not the same as any other. Because the duration of time-units which follow each other in a non-recurrent sequence recurs, one can compare, by reference to them, successive happenings of other sequences with each other, in terms of their duration – which would be impossible without a graduated reference movement passing equal spaces at equal speed and which was, in fact, not possible for people as long as they had not learned to use or to produce on their own movements with these characteristics as timing devices.

Moreover, by varying the speed of movement of the timing device, people can vary the length of the movement's passage from an earlier to a later position – the greater the speed, the shorter the passage through equidistant intervals; and by combining movements of different speeds into one mechanical unit people can arrange them in such a way that the speed of any faster moving unit stands in a set proportion to that of any slower moving unit. We say, in our process-reducing language, that a day has twenty-four hours and an hour sixty minutes, each of which has sixty seconds. If one is able to look at a watch for a while, unhampered by this socially standardized conceptualization and the mode of experience that goes with it, one can easily recognize that one looks at sets of physical units moving at different speeds so that the length of their passage through the same interval differs in precisely set proportions. Thus, if one imagines a clock with a date indicator moving in the same way as the other hands, that is a day-hand in addition to those indicating hours, minutes and seconds, one can take one's choice – one may see together the four moving hands as indicators of four interrelated types of time intervals called days, hours, minutes and seconds, or simply as four different physical units moving around at different speeds in the set proportion of 1:2:24:1440. In accordance with the conceptual tradition of our societies, we read and experience the changing configurations of these moving units on the face of a watch in terms such as 'five minutes past seven' or 'ten minutes and thirty-five seconds'. In that way, moving configurations used for timing events are transformed by the social customs of the beholders into symbols of instances in the flux of incorporeal 'time' which, according to a common use of the term, appears to run its course independently of both any physical movement and any human beholder.

The puzzle of 'time', the use of the term as if 'time' had an independent existence, is certainly a striking example of the way in which a widely used symbol, cut loose from any observable data, in common discourse can assume a life of its own. What has been said so far may go some way towards explaining the impression of an independent existence of 'time'. This impression, as I have already said, is connected with the fact that 'time', in common with a whole set of other social institutions, is relatively independent of any particular human being, though not of human beings in their capacity as societies or humankind. It is also connected with the related fact that timing devices, human-made or not, are self-moving; they are socially standardized movements which have a measure of independence of other movements and of other changes generally, non-human or human, for which they may serve as gauges.

It is easy enough to observe in our social life this double-layered relative independence of 'time' indicated by clocks – its relative independence as a social institution and as an aspect of a physical movement. Just as languages can only fulfil their function as long as they are common languages of whole groups of people and would lose it if every person made up a language of his own, so clocks can fulfil their function only if the changing configuration of their moving hands, in short the 'time' they indicate, is the same for whole groups of people; they would lose their function as timing devices if every person made up a 'time' of his own. This is one of the sources of the compelling force which 'time' has in relation to an individual person. He has to attune his own conduct to the established 'time' of any group of which he forms part and, as one shall see, the longer, the more differentiated become the chains of functional interdependencies which bind people to each other, the stricter becomes the regimen of the clocks.

25 Like many other social skills, timing has grown into its present condition slowly over the centuries in reciprocal conjunction with the growth of specific social requirements. Foremost among them is the need for people to co-ordinate, to synchronize – their own activities with each

other and with the succession of non-human natural events. Such a need does not exist in all human societies. It makes itself felt more and more strongly as the societies formed by people become larger, more populous and more complex. In early groups of hunters, herdsmen and tillers of the land the need for actively timing or dating events is minimal and so are the means of so doing. In large urbanized state societies, above all, in those where the specialization of social functions is far advanced, where the chains of interdependencies binding the performers of these functions to each other are long and highly diversified and where much of people's daily toil has been taken over by human-made energies and machines, the need for timing and the means of satisfying it, the signals of mechanical time-pieces, become inescapable and so, therefore, does people's sense of time.

As the ancestors of all members of machine-societies with their innumerable watches and clocks lived in small groups as hunters, herdsmen or agriculturalists without such devices, it betrays a somewhat naive self-centredness if members of these complex societies – and especially their highly respected spokesmen in such matters, their philosophers – discuss the problems of 'time' on the tacit assumption that people's concept and experience of time has been throughout the ages the same, namely their own. In that respect philosophers simply follow the usage of their society. They accept at face value the concept of time which they have learned from their elders while growing up and which is used routinely in their society by everyone day by day; but they do not ask how and why the experience of time has attained such ascendancy. They receive the concept and institution of time as a gift, as part of the common framework of symbols used in their society, as a means of both communication and orientation, but they do not explain it – they do not ask which sequence of changes in the manner of living and people's experience went into its making. In the philosophers' perspective the concept of 'time' seems to stand on its own, though linked to that of space. Hence 'time' itself seems to stand on its own: a separate word, perhaps fortified by an isolating definition, seems to indicate a separate existence. To pursue it has seemed the task of philosophers.

This has been – down the centuries – a wild-goose chase, a hunt for something that does not exist, a hunt for 'time' as an immutable datum given to and experienced by all humans in the same manner. By framing their problems in terms of this 'object', philosophers have been caught in a self-made trap. They have been confronted again and again by a choice between two irreconcilable basic assumptions about 'time', both equally speculative and untestable. They have on the one hand put forward the assumption that 'time' is given to people like any other physical object as part of the eternal order of nature. This alternative was Newton's choice. On the other hand, philosophers have come up with the assumption that 'time' is a universal form of human consciousness; they have argued that humans can and must perform the synthesis of events in terms of time always and everywhere in the same way, without any learning and prior to any experience of objects. In that view, 'time' as a way of ordering events is 'laid on' in people – on its own or in conjunction with 'space' – as part of their power to reason, as an immutable property of the human mind or of human existence.

The two polar views about the nature of 'time' just sketched are in agreement with the basic assumption of classical European philosophy about the acquisition of knowledge. This assumption which, at least since the time of Descartes, all the various epistemological views of philosophers have in common, again remains unexamined. The common question which philosophers think they must answer is the question as to whether or how far the knowledge 'inside' people corresponds to the objects 'outside', often conceptualized by philosophers as the 'external world'. This way of framing the cognitive problem with its nightmarish imagery of an invisible gulf, a quasi-spatial divide, between individuals perceived as closed containers holding within them knowledge, while the whole world exists 'outside' of them, has dominated philosophical arguments for centuries and it is time to make an end of it.

The assumption of this quasi-spatial divide between an 'inner world' and an 'external world' is an individual and collective fantasy. It symbolically represents a conceptual reification of a specific type of experience. People who are engaged in a process of solving scientific problems (or, for that matter, any problem requiring, as is the rule in complex societies, a relatively high degree of emotional restraint, a relatively long pause of reflection between an impulse to act and its execution), have to distance themselves in their imagination again and again from the object of their reflection. In reflecting about the peculiarity of scientific knowledge philosophers have inadvertently reified the act of distancing performed by a scientist. These acts of distancing are imperceptibly transformed into the notion of a real distance between themselves, the subject, and that which presents a so far unsolved problem which the perceiving subject tries to solve – his 'object'. In the philosophers' reflections this experience of distancing presents itself as an invisible gulf between 'subject' and 'object', 'inner' and 'outer' worlds. Misled by their own figurative use of spatial metaphors to describe the functions of human-made symbols – of knowledge – within a process of problem-solving, they attribute to the unsolved problem, the 'object' of their reflection, a position in space 'outside' of themselves. Two concepts, 'subject' and 'object', which simply refer to two inseparable complementary functions of humans and nature or, for that matter, of humans and humans *in a process of cognition*, are made to appear, in philosophical discourse, as two independent existences separated from each other by an invisible spatial divide. The world, in philosophical parlance, is 'external' and knowledge 'internal'. But 'knowledge', like 'speech', presupposes a plurality of communicating people, not just an individual. 'The external world' would scarcely be an apt means of conceptualizing the relation of multipersonal concepts to their unipersonal use. The 'object' is a function of the fund of social knowledge existing at the time.

Given this common layout of the philosophers' problem of cognition and knowledge, the perimeter of possible solutions is set between two polar views: either the 'external' objects project their image into people whose knowledge is the result of this projection; or people project specific experiential forms of their 'internal' intellectual constitution, their unchanging 'reason', 'mind', 'consciousness', 'existence' or whatever, with its unalterable laws or categories, on to the objects of the external world. The first of these alternatives tends towards 'naive' positivism, the second towards nominalism and solipsism. According to the latter view, the world is basically unknowable as all experiences of this 'external' world are ordered in accordance with the predetermined properties of an individual's own intellect. They are supposed to be the same in all human beings (which is wholly inconsistent with the premises of the phenomenalist argument, for 'other human beings', on this assumption, are just as 'external' to oneself as inanimate objects and any general statement about people valid for all, becomes, therefore, impossible). In essence, this view represents the credo of a solipsist: each person alone in an unknowable world. Needless to say, this view itself, like that of its polar opponents, is put forward as true knowledge about humans and their knowledge. A long procession of books written on these lines, a tragi-comic masquerade of wasted lives, litters humankind's trail. If the world 'an sich' is unknowable, one wonders why their authors bother, often rather emphatically, to state their case. Resigned silence might be more appropriate.

The polar views on time are set in the same mould. At one end of the spectrum is the conception of time as a property of the 'objects', at the other, as a property of the 'subjects' of knowledge. The antagonists have in common the assumption that their own concept and experience of time is a universal concept which they share with all human beings regardless of the condition of their knowledge and, generally, their manner of life. Being philosophers and thus having, at best, a limited knowledge of societies other than their own, their imagination flags. Their own synthesis of events in terms of time appeared as inescapably that of all people. They could not imagine that the type and range of the synthesis performed by human beings knowing less and living differently may have been different from their own.

26 The philosophers' approach to the problem of time is thus somewhat lopsided and infected by heteronomous valuations. It is onesided because they cannot rid themselves of the notion that people always and everywhere seek to order the course of events by the same concept of 'time' which they use themselves. They apply a onesided concept of time much as if the concept of the 'organism' were developed solely from the example of humans. The philosophers' time concept is further influenced by extraneous valuations; for their cognitive effort is directed solely at the discovery of 'time' as an unchanging universal 'beyond time'. Their aim reflects the fundamental if unacknowledged value scheme which they have inherited from theology (where it makes more sense). They seek to discover 'time' as something beyond, and apart from, the observable succession of changes, from the 'flow of time'

itself – as an unchanging and eternal property of either nature or people. With very few exceptions, such as Hegel or Comte, philosophers have succumbed again and again to the temptation to reduce observable processes as inessential to something absolutely immutable. The concept of time refers among other things to unrepeatable processes. An eighty-year-old will never again be forty. The year 1982 will never return. But the great majority of philosophers have sought over and again for something absolutely unchanging and stationary behind all changes, such as an eternally constant consciousness of time or space in people, or eternal laws of nature or reason.

In Galileo's hands this process-reducing method of enquiry – this 'scientific method' – still had largely an instrumental character; it was a means to an end. The mathematical formula expressing certain observable regularities of falling bodies never entirely lost for Galileo its significance as a rule of thumb to be used for practical tasks. It was not yet sanctified by the name of 'law' – the 'law of falling bodies'. In course of time, however, Galileo's procedure, applied at first only to specialized and limited tasks, was developed further; it was applied above all to the exploration of regularities of the heavenly bodies. Increasingly it assumed, in the understanding of its specialized practitioners and the explicit interpretation of philosophers, the character of an eternal and exclusive method of discovery derived either from people's innate 'rationality' or the nature of their objects. From being a means to an end, therefore, the discovery of unchanging regularities based on systematic measurements and represented by mathematical symbols of eternity such as the 'laws' of reason, of logic or the results of pure mathematical operations, the 'laws' of nature, became for a while the focal point of physical and philosophical orthodoxy and their discovery, accordingly, came to be regarded as the highest, the most prestigious goal of scientific work.

Thus universalized and ritualized, this discovery of something eternally unchanging behind all temporal changes as the highest reward of people's quest for knowledge – not only in physics but also, in proportion to their prestige, in many other fields of enquiry – gained and retained its regard among people partly because it provided them, up to a point, with better means of orientation and control than they had had before, and partly because it agreed with a traditional value scheme which had its roots in quite different aims than those of orientation and control; it sprang not from people's quest for knowledge as such, but from that for something quite extraneous to this quest, namely from their longing for something permanent behind the impermanence of all observable data, from their quest for something imperishable and timeless as the solid fundament of their transient lives. The high value attached to these cognitive symbols of permanence came to be heteronomously superimposed upon the intrinsic and instrumental cognitive value of the symbolic representation of a sequence of changes by means of something unchanging.

27 In the case of Galileo's symbolic representation of an accelerating movement through an unchanging acceleration constant, that is his representation of a sequence of changes by means of a changeless mathematical formula, the instrumental cognitive function of this process-reducing method still clearly prevailed. Since then routinization and a constant flow of philosophical indoctrination have made it increasingly difficult to notice any longer the peculiar flavour of the term 'law' used figuratively as symbol of an invisible changeless order underlying the visible world of continuous changes in the midst of which we live and of which we form part.

The adequacy of this process-reduction, this symbolic representation of sequences of change through unchanging laws and, more generally, law-like abstractions as *purely instrumental* cognitive devices can be tested; these devices themselves, therefore, are potentially changeable and replaceable by more adequate instruments of research. The development of scientific practice has already led in a number of fields to a change in the function attributed to timeless laws and law-like devices; their use has become more limited and their cognitive status as the highest aim of discovery less assured than it used to be in the age of Newton. In many sciences, including some branches of physics, models of processes, of long sequences of change 'in the succession of time', such as cosmic and biological evolution or social development are in the ascendancy. But at present the recognition of the cognitive function of process-models as focal points of scientific theories is hampered because they fail to live up to the expectations which 'laws' and law-centred theories appeared to fulfil. They do not direct the effort of cognition at something imperishable behind the transient world and therefore fail to fulfil the hope that science will discover that which transcends all changes – which is timeless and imperishable. Instead they direct the effort of cognition at the progressive discovery of change-immanent structures and regularities, of the order of changes in the sequence of time itself. Hence the extra-scientific semi-religious function which gave – and still gives – to concepts such as 'laws of nature', 'order of nature' or 'laws of reasoning' a social status far beyond their instrinsic instrumental cognitive value, is lost.

The personal reason why the discovery of that which is eternal and permanent behind all changes has a high value for people is, I suggest, their fear of their own transcience – the fear of death. Once it was overcome by the thought of the eternal gods and then, again, people tried to shield themselves against it by the thought of the eternal laws of nature representing nature's imperishable order.

The special warmth with which Kant in a well-known passage spoke of the eternal laws of the movements in the heavens and the eternal moral law in our breasts is one example of the emotional significance of what appear on the face of it as purely 'rational' scientific or philosophical ideas. The persistence with which the discovery of eternal laws or law-like connections beyond time is proclaimed as the highest goal of enquiries that can claim to be scientific is another, as is the high cognitive value attached by some philosophers to formal logic or pure mathematics.

A few examples taken from G. H. Hardy's well-known short book *A Mathematician's Apology*[15] may help to illustrate my point. He states there, very understandably, that 'the noblest ambition is that of leaving behind one something of permanent value'. He goes on to praise mathematics because it is capable of fulfilling that ambition better than most other fields of human endeavour. 'Greek mathematics is permanent, more permanent than Greek literature ... Archimedes will be remembered when Aeschylus is forgotten ... Mathematical fame, if you have the cash to pay for it, is one of the soundest and steadiest investments.' He goes on to report Bertrand Russell's dream which, indeed, ought not to be forgotten:

Yet how painful it is to feel that with all these advantages, one may fail. I can remember Bertrand Russell telling me of a horrible dream. He was in the top floor of the University Library about A.D. 2100. A library assistant was going round the shelves carrying an enormous bucket, taking down book after book, glancing at them, restoring them to the shelves or dumping them into the bucket. At last he came to three large volumes which Russell could recognise as the last surviving copy of *Principia Mathematica*. He took down one of the volumes, turned over a few pages, seemed puzzled for a moment by the curious symbolism, closed the volume, balanced it in his hand and hesitated ...

28 What are people not willing to believe in order to conceal or sweeten the idea of the finitude of their lives, the thought of their deaths! The high status of mathematics in our societies undoubtedly rests, among other things, on the fact that it is one of the symbolic structures in whose name one may claim, as Hardy does, to offer eternal verities outlasting death.

If the claim were somewhat more modest, one might concur with it. It can certainly be of value for people to create something that may have a meaning and function for future generations. But Hardy succumbs to the temptation of according to the manipulation of human-made symbols of relations, to the discovery of their immanent order, in which specialists in pure mathematics engage, an unusually high chance of achieving immortality. As far as immortality is concerned, so he puts it, one can scarcely invest one's energies in a better field than mathematics. And he goes as far as to assure his readers that the name of a mathematician like Archimedes will still be familiar to all people when that of a Greek dramatist like Aeschylus is long forgotten. He unquestioningly presupposes, in other words, that future generations will and must share his own scale of values, which is perhaps not unrepresentative of his society and in which the eternal truths of mathematics rank higher than fragile works of art.

Such ideas certainly flatter people's insatiable desire to hear of something that can appease the thought of their own transience. But this status-raising role of mathematics as guarantor of permanence stands on uncertain ground. This is obscured by a gap in the present fund of knowledge. Just as, up to now, a generally accepted theory of time capable of being tested and developed has been lacking, so too is a generally accepted theory of mathematics, a theory of mathematics as a branch of knowledge. Moreover, since this discipline is finally human-made and so a society-related field, what is lacking is a theory of mathematics in terms of the sociology of knowledge, a theory able to explain, among other things, the functions of mathematics, its usefulness in solving physical problems and its applicability to natural connections.

It is enough here to point to these gaps of knowledge in passing and to add a word on the direction in which one might seek something to fill them. To do this one needs a theory of social symbols. I have already referred to their realm by the concept of the fifth dimension. Humans are figures in time and space and can at all times be located and dated in terms of their position within these four dimensions. But that is not enough. The fifth co-ordinate of people and all they experience and do is that which marks their passage through the symbolic universe in which they live together. An obvious representative of this dimension is language, comprehensive, complex human-made symbols which may differ from society to society and which serve both communication between people and their orientation. But symbolic contents such as concepts or what we call the 'meaning' of communications – in short, everything which in human intercourse passes through and is shaped by 'consciousness' – are also part of this dimension, including, without doubt, the present meanings of the concepts 'space' and 'time'. These, like other human-made symbols are not simply there, once and for all. They are always in flux, always having become what they are and always evolving; they evolve in one direction or another, either towards greater approximation to reality and object-adequacy, or towards an intensified character as expressions of human affects and fantasies, or, again, they evolve towards a greater or lesser degree of synthesis.

29 The development of the standard of human activities and ideas within the framework of what we understand today as 'time', is itself a good example of a gradually extending synthesis. An expression like 'When we feel

cold' is characteristic of the kind of time-definition prevailing in societies at an earlier stage of development. At a somewhat later stage a human group may already possess the less personal symbol 'winter'. Today a calendar is used throughout the world which shows in which month winter begins; and this calendar is used in parts of the world where the 'winter' months are decidedly warm.

It is unlikely that the people of all earlier stages of society found it necessary to engage in what we today call 'timing'. But if one follows the early traces of such activities, it is not difficult to see that those people were always concerned with very personal connections between themselves and a specific, visible or tangible entity. For example, people experienced the appearance of what we call the 'sun' or 'new moon' as a sign that they should or should not do a particular thing. However, it is a peculiarity of the concept of 'time' not only that it is a symbol of an extremely extensive synthesis, of a very high-level 'abstraction', but also that, while being a symbol of relationships, it does not symbolize relationships between particular persons or situations. In this respect time belongs to the same order of symbols as those with which mathematicians work. It is a purely relational symbol. While 'time' is itself a symbol of relations of a specific kind, for example, between positions in two sequences of events, the events related in this way are interchangeable. Identity of relationships is compatible with variation of what it related.

Humankind had to cover a long distance before people were in a position, and found it necessary, to create symbols of pure relationship. But although the formation of such symbols presupposes a capacity for relatively extensive syntheses or, in familiar language, a highly developed capacity for abstraction, nevertheless the most elementary statements that can be made about it are simple enough.

If twice two apples are placed side by side, that makes four apples. There are stages in the development of society at which people possess symbols for 'four apples', 'four cows' etc., but not yet symbols such as 'four', 'five', 'six', referring to no specific objects and for that reason capable of being applied to a multiplicity of different objects. Here, then, is the key to the secret of the applicability of mathematics to so many different areas. In all these areas there are specific relationships. By means of measurements these relationships can be represented by pure mathematical relation-symbols. Pure relation-symbols can be manipulated, for example on paper, quite differently to relations between specific objects or persons. But the result of such purely symbolic manipulations can then be applied back to relations between specific objects or persons. And one can perhaps determine experimentally whether the results calculated by the manipulaton of pure relation-symbols are confirmed or not by being referred back to specific relationships.

'Timing', as I have said, is an activity in which people relate together succession-aspects of at least two sequences of events, one of which is socially standardized by others as a meter of intervals or positions in the succession of events. The position in the sequence of socially standardized time-units that we represent as 'twelve-thirty p.m.' can serve as a reference point for a multitude of different sequences of actions and events. It can refer just as well to the departure of a train as to the beginning of a solar eclipse or the end of a school lesson.

Perhaps the little that it has been possible to say here about pure relation-symbols is enough to make it clear that symbols which presuppose a synthesis as comprehensive as this represent a relatively late stage in the development of human symbols and of the corresponding social institutions. A theory of the development of human symbols would be needed in order to gain a firmer grasp of such problems. Until this gap in the existing fund of knowledge is closed a large number of problems will remain insoluble. The problem of time is one of them.

I once read the story of a group of people who climbed higher and higher in an unknown and very high tower. The first generation got as far as the fifth storey, the second reached the seventh, the third the tenth. In the course of time their descendants attained the hundredth storey. Then the stairs gave way. The people established themselves on the hundredth storey. With the passage of time they forgot that their ancestors had ever lived on lower floors and how they had arrived at the hundredth floor. They saw the world and themselves from the perspective of the hundredth floor, without knowing how people had arrived there. They even regarded the ideas they formed from the perspective of their floor as universal human ideas.

The vain attempt to solve a problem as fundamentally simple as that of time is a good example of the consequences of forgetting the social past. By remembering it, one discovers oneself.

30 All the human beings who form with each other one of the highly differentiated industrial nation states of our age, have ancestors who, at some time in the past, formed with each other tribal groups and perhaps village states characteristic of the same stage of development represented today, for instance, by some of the Amazon Indian tribes. Like the latter, the former too, however different these nation states may be from each other, share with each other specific personality characteristics as representatives of the same stage in the development of societies. A specific time experience is one of these common characteristics. Members of industrial nation states usually have an almost inescapable need to know, at least approximately, what time it is. This need, this all-pervasive sense of time,

is so compelling that most members of these societies find it extremely difficult, if not impossible, to imagine that their own time experience is not shared by human beings everywhere. This sense of time seems so deep-rooted, so much forms part of people's personalities, that they find it hard to see in it a result of social experiences. There is a widespread tendency among members of these societies to regard as entirely their own only what they perceive as a gift of nature or perhaps of the gods. What is socially acquired, their social habitus, often appears to them as accidental in relation to their true nature, as a façade that may be easily removed.

The compelling character of the sense of time, as normally experienced by members of more highly differentiated societies, can help to correct that view. This time experience very much forms part of what people in these societies experience as their own self. They may hate the inner voice which, in Auden's words, 'coughs if one would kiss', and yet they cannot get rid of it. It is one of the civilizing restraints which, though not part of human nature, are made possible by it. They form part of what is called second nature, of the social habitus which is a characteristic of the individuality of every human being.

What may not be always quite clear, what does not yet form part of the common knowledge of the age, is the fact that differences in the social habitus of members of different societies are often responsible for difficulties or even for blockages of the communication between them. Such blockages are most likely to occur and also most severe if societies coming in contact with each other represent different stages of social development. At present, communication difficulties or blockages of this kind are often rendered more unmanageable than they inherently are by the use of diagnostic terms which are imprecise and equivocal. After the term 'racial differences' has been abandoned one has found refuge in such expressions as 'ethnic differences' which avoid the need to say clearly whether the differences to which one refers are more genetically determined or more socially acquired. Differences in the time experience of different societies, like other aspects of civilizing processes, allow in that respect an unambiguous decision. These differences are, without doubt, socially acquired. They are characteristic of differences in the social habitus and thus in the personality structure of people belonging to different societies – differences which are most pronounced when these societies represent different stages of social development. As social differences they are amenable to change if that is necessary, though it may be a slow change. A three-generation model may be needed in order to realize such a change.

An experience from the 1930s told by Edward T. Hall[16] vividly illustrates the contrasting time experience of Americans and Pueblo Indians. Hall refers to the great strictness of Americans in matters of time. Lack of it can easily appear either as insult or as irresponsibility. There are, he observes, extreme cases where people become time-ridden, obsessed by the need not to 'waste time', to be everywhere 'on time'. The occasion of the clash between the sense of time of Americans and of Pueblo Indians was a Christmas dance of the latter visited by groups of the former. The Indians lived high up near the Rio Grande:

At seven thousand feet, the ordeal of winter cold at one o'clock in the morning is almost unbearable. Shivering in the still darkness of the pueblo, I kept searching for a clue as to when the dance would begin.
Outside everything was impenetrably quiet. Occasionally there was the muffled beat of a deep pueblo drum, the opening of a door, or the piercing of the night's darkness with a shaft of light. In the church where the dance was to take place a few white townsfolk were huddled together on a balcony, groping for some clue which would suggest how much longer they were going to suffer. 'Last year I heard they started at ten o'clock'.

'They can't start until the priest comes.' 'There is no way of telling when they will start.' All this punctuated by chattering teeth and the stamping of feet to keep up circulation.

Suddenly an Indian opened the door, entered, and poked up the fire in the stove. Everyone nudged his neighbour: 'Maybe they are going to begin now.' Another hour passed. Another Indian came in from outside, walked across the nave of the church, and disappeared through another door, 'Certainly now they will begin. After all, it's almost two o'clock.' Someone guessed that they were just being ornery in the hope that the white men would go away. Another had a friend in the pueblo and went to his house to ask when the dance would begin. Nobody knew. Suddenly when the whites were almost exhausted, there burst upon the night the deep sounds of the drums, rattles, and low male voices singing. Without warning the dance had begun.

For Pueblo Indians, in accordance with their tradition, a dance of this kind, whatever its other functions might have been, always had a ritual function as well. Traditionally a dance of this kind was experienced as a communication and perhaps an identification with ancestral spirits or, at any rate, with the spirit world. The participants started dancing when they were in the right mood. Their traditional way of life required time discipline if at all, only on a few occasions, for example, for the provision of food. But the discipline was imposed by tangible pressures such as that of hunger or the anticipation of hunger. It was not time as a voice of their individual conscience which called them to task. In the course of time, no doubt, the ritual significance of the dancing became weaker and, in some cases, its financial significance stronger. The dancing itself, if it did not lapse, changed its character; it assumed the form of an actors' performance. But such a change of tradition and the transformation of the personality structure which it required was a difficult process involving at least three generations and was often very painful. Once more an example may help.

The superintendent of a Sioux reservation made the following remark when discussing the adjustment difficulties of the tribal groups. He himself was a man of mixed descent, who, as a child had lived in the reservation. He had then apparently a normal American upbringing, had studied at an American university and taken a degree. When he became superintendent of a Sioux reservation he had, understandably, the American sense of time and was at a loss to understand why his charges, the Sioux, did not have the same time discipline which he had:

'What would you think of people,' he said, 'who had no word for time? My people have no word for "late" or for "waiting", for that matter. They don't know what it is to wait or to be late.' He then continued, 'I decided that until they could tell time and knew what time was they could never adjust themselves to white culture. So I set about to teach them time. There wasn't a clock that was running in any of the reservation classrooms. So I first bought some decent clocks. Then I made the school buses start on time, and if an Indian was two minutes late that was just too bad. The bus started at eight forty-two and he had to be there.'[17]

A child growing up in one of the highly time-regulated and industrialized state societies of the twentieth century needs from seven to nine years to 'learn time', that is to read accurately, to understand the intricate symbolism of watches, clocks and calendars and to regulate his own feeling and conduct accordingly. Once they have learnt it, however, members of these societies appear to forget that they had to learn time. It is completely obvious to them that one regulates one's day and nights directly or indirectly in terms of the time signals brought to one's notice by one or another of the technical devices whose function it is to do that. The self-control agencies of a person, 'reason' and 'conscience' or however one might call them, are formed accordingly. They are heavily reinforced by social constraint operating in the same direction. Human relationships of all kinds, in societies at that stage, would be severely disturbed and, in the long run, could hardly be maintained if one ceased to regulate one's own conduct in terms of a communal time-schedule.

Properties of one's own person which are so inescapable and compelling, which are moreover shared by most other people one knows, are often conceptualized and experienced, in accordance with the dominant code of knowledge, as natural properties, as attributes with which one is born. Hence, most people who have been brought up in one of these highly time-regulated societies react like the superintendent of the Sioux reservation if they are confronted by human beings who are not time-conscious and time-regulated in the same way as they are. They can hardly believe that human beings exist who are not as time-regulated as they are and who may not even have a word for 'time'.

The difficulty is that the highly time-regulated members of later societies not only fail to understand members of earlier societies with few timing needs. They also fail to understand themselves. The categorical equipment at their disposal offers one main conceptual device for diagnosing and explaining so compelling and inescapable an attribute of their own person as their time-experience. That is the diagnosis and explanation as an unlearned characteristic of human nature, perhaps in a disguised form as a conceptual synthesis prior to all experience. It is difficult to overlook that timing has to be learned. Yet, once acquired the ever-present sense of time is so compelling that it appears to be part of one's natural make-up. There is as yet no clear and certainly no universal understanding of the fact that a learned, that is socially acquired patterning of the natural organization of humans can be almost as compelling and inescapable as the genetically determined structure of a person. The time-experience of people who belong to firmly time-regulated societies is one of many examples of personality structures which are as compelling as biological characteristics, yet socially acquired. This accounts for the seemingly self-evident expectation of those brought up in highly differentiated societies that their own time experience is a universal gift of all human beings and for the disbelief or the surprise with which they often react if they encounter, or hear of, human beings of societies which are not time-regulated in the same manner.

31 One cannot entirely avoid referring in this context to the present condition of the science of psychology. It would not be unreasonable to think that this science could be of help in elucidating the differences in the time experience and in the time regulation of conduct of members of different societies. However, as presently taught at academic institutions, psychology is of little assistance here. There are good reasons which help to explain why that is the case. Many dominant schools of academic psychology appear to share the conviction that a sharp distinguishing line can be drawn between psychology proper and social psychology. The distinction is based on an assumption which has often the character and the force of a seemingly self-evident axiom. It plays a decisive part in the traditional layout and the procedure of a number of human sciences. The assumption is that the scientific study of individuals and that of societies can be pursued independently of each other, as it were in separate compartments.

The institutional division between psychology and social psychology is an example. It is justified only as long as one adheres to the traditional belief that some aspects of the psychological make-up of human beings are purely 'individual' and totally independent of the fact that a person grows up and normally lives among other persons, while other aspects are purely 'social' and as such can be divorced from the 'individual' aspects. Extraneous, in particular political, values and ideals evidently play a decisive part in keeping alive a notion so obviously erroneous as that of a quasi-ontological separation of 'individual' and 'society'. In order to persist in this traditional misconception, one had to close one's eyes to the obvious fact that a human child, in order to become fully human, has to learn a social language which the child has in common with other human beings, which transmits to the child and, again, from the child a good deal of group-specific knowledge which, in short, enables the child to communicate with other human beings. The language absorbed by a child together with other experiences forms part of the early social layers of an individual's personality structure. Individual properties characteristic of the distinctness from any other do not develop as it were independently of and in separation from these social characteristics. The individual cast of speech or, for that matter, of writing does not emerge outside; it develops as a unique patterning of the common speech and writing-form of a society. A social habitus, in other words, forms an integral part of the unique personality structure of every human individual.

The institutional separation between individual and social psychology blocks the recognition of the inseparability of shared social and unique individual personality structures in a human person. It has led to the attempt made by psychologists to present their own field as a natural science and to adopt methods of research befitting a science of that type. Thus one is confronted by a somewhat odd situation; individual psychology, it appears, is a natural science, social psychology a social science.

In actual fact the division is untenable. Every person tested by individual psychologists in their experiments, has fashioned in a personal way from early childhood on, what he has learned from others, has in common with others, has experienced in relation to others. Thus, a common social heritage, above all a heritage of verbal and other social symbols, assumes in the individual person a unique shape, an individuality which differs more or less from that of every other member of the same society. Yet, although the psychological levels of a human person, whether conduct or feeling, conscience or drive, are invariably patterned by learning and thus have natural and social characteristics at the same time, quite a number of individual psychologists proceed in their research as if the persons they study were natural objects pure and simple unaffected by their social language or any other social patterning. A physiologist investigating with the help of a number of individual human beings physiological functions of the human person has a fair chance to discover regularities of human nature common to all human beings regardless of the stage of development of their society. If psychologists make the same experimental arrangement, their results will be suspect and probably invalid; for the psychological levels of human beings are not only structured by nature or, in other words, characteristic of the whole human species, they are also structured socially and are thus group-specific as well. It is significant that so far technical terms referring to the social aspects of an individual person have not gained wide currency in the languages of our time, although the recognition of psychology as a social science is not new. I am well aware of it.

The examples I have given may be of help in demonstrating and in making accessible to one's own experience what is meant if one refers to the *social habitus* or the *social personality structure* of individual people. The almost inexorable self-regulation in terms of time of people brought up in highly time-regulated societies is a good example of what is meant if one speaks of the social habitus of an individual person. It would be easy to expand what has been said before by pointing out that the experience of past, present and future also differs in societies at different stages of social development. Just as the chains of interdependency in the case of pre-state societies are comparatively short, so their members' experience of past and future as distinct from the present is less developed. In people's experience, the immediate present – that which is here and now – stands out more sharply than either past or future. Human actions, too, tend to be more highly centred on present needs and impulses. In later societies, on the other hand, past, present and future are more sharply distinguished. The need and the capacity to foresee, and thus considerations of a relatively distant future, gain stronger and stronger influence on all activities to be undertaken here and now.

32 Although in this context social differentials of timing and time experiences command special attention, I have been anxious not to give the impression that time experiences as an ingredient of the social habitus of people can be studied in isolation. I have already referred to some other aspects of people's social habitus which can be used as criteria of different stages in the development of societies, among them differences in the level of conceptual synthesis and in the restraint characteristics of gods. The stage-specific character of people's sense of the future shows once more the close links between time experience and civilization. Acting more in terms of present needs than in terms of a future requires less – and less even – self-restraint; acting and planning in terms of a future, perhaps even a relatively distant future requires a capacity for subordinating present needs to expected future rewards. People brought up in societies with a future centred code may take the corresponding pattern of self-regulation for granted; they may even consider it as a normal human propensity. They may not be aware that this pattern of self-regulation and, as an aspect of it, a highly differen-

tiated regulation in terms of time, like other social skills, has grown into its present condition slowly over the centuries in conjunction with the growth of specific social requirements.

I have indicated some of them. One had to attune all one's activities to those of a growing number of people, had to perform one's own activities, including those of getting up and going to bed, more and more precisely at such and such a time, had to plan ahead more and more accurately when in the future one needed or wanted to do this or that. Thus, the self-regulation of people, both 'social' and 'individual', grew to its present highly differentiated condition in connection with corresponding changes in the structure of human societies or, in other words, with changes in the figuration people formed with each other. A striking, if intermittent, increase in the number of people on earth, going sometimes hand in hand with equally striking advances of occupational specialization and organizational integration, resulted, as I have indicated elsewhere,[18] in an even greater increase in possible human relationships. Chains of interdependencies between people became not only longer, but also more differentiated; their web became more complex and the accurate timing of all relationshps became more pressing, and in fact indispensable, as a means of regulating them. Thus the highly differentiated and unrelenting attention to time which forms part of the social habitus of those brought up in more differentiated and more complex industrial state societies is no more surprising then the capacity of hunting tribesmen to form a detailed picture of their prey from a couple of footprints.

The personality structure of humans develops along different lines in accordance with differences in the structure and thus with the developmental stage of the societies where they grew up. If they belong to relatively small undifferentiated societies with only a few point-like and intermittent timing requirements, their self-steering and restraint patterns develop accordingly. If they belong to one of the large, populous and differentiated industrial societies with a correspondingly differentiated and continuous time-regulation, again, the self-regulation and, in a wider sense, the social habitus of individuals will develop in accordance with structural characteristics of these societies. Once more one can notice here connections linking the development of timing as a social skill and a regulator of human feeling and conduct, with the development of civilizing restraints. The connection may be of help in clarifying a central aspect of civilizing processes to which I have referred before and which is often misunderstood.

There is no zero-point of civilizing processes, no point at which human beings are uncivilized and as it were begin to be civilized. No human being lacks the capacity for self-restraint. No human group could function for any length of time whose adults failed to develop, within the wild and at first totally unrestrained little beings, as which humans are born, patterns of self-regulation and self-restrained. What changes in the course of a civilizing process are the social patterns of individual self-restraint and the manner in which they are built into the individual person in the form of what one now calls 'conscience' or perhaps 'reason'. In earlier societies which are relatively small, more autarchic and less differentiated, the social code of a group may demand forms of restraint which are geared to specific occasions; they may even be according to the standards of later societies extremely severe forms of self-restraint.

Initiation rites can include extremely frightening experiences, which may result in a lasting submission to specific taboos, in a fear of transgression in a specific area. But it may also leave the way free to emotionally charged activities on other occasions, to a free flow of passion of a strength and intensity no longer accessible to people at a later stage of development and confined, like the intensive self-restraint, to specific socially prescribed occasions. Even in Medieval societies, which in developmental terms were considerably more differentiated and more complex than tribal societies, contrasts and fluctuations in the regulative pattern of the social code were a normal feature of people's lives. Indulgence in wild pleasures might be followed by forms of self-tormenting repentance. Fasts followed the carnival. Extreme forms of asceticism in some monkish orders sometimes existed side by side with a very free indulgence in the pleasures of life. During the earlier stages of a civilizing process, one might say, conscience-formation tends to be patchy, extremely strong and severe in some respects or on some occasions, extremely weak and lenient in other respects. Characteristic of the restraint-pattern of civilizing processes at the later stages is the tendency to be temperate and even in almost all respects, on almost all occasions. The time-regulation characteristic of these societies is, as one may see, representative of their civilizing pattern. The time regulation, too, is no longer point-like and patchy. It covers the whole life of people. It admits of no fluctuations; it is even and quite inescapable.

It is difficult to escape the tendency to simplify distortions. In this case one of the most persistent is the idea that stages in a civilizing process can be most easily determined in purely quantitative terms. A certain inadequacy of the present conceptual equipment, moreover, has given support to this tendency. In order to avoid representing a civilizing process in terms of static polarities such as 'uncivilized' and 'civilized' one has to resort to such terms as more civilized or less civilized which can easily suggest a rise or decline in the quantity of self-restraint. Following this line of thought one may come to believe that human beings in the early stages of their social development live together with a small quantity of, if not entirely without, social and individual patterns of self-regulation and self-restraint. One may imagine that all that changed in the course of the process of civilization was, as it were, the quantity of self-restraint. What has been said here about the development of timing as a social skill and a social regulator may help towards a better understanding of the non-quantitative aspects of change characteristic of a civilizing process. Once more an inadequacy of present day languages obtrudes itself on the way of thinking. If I say 'non-quantitative' some readers will almost automatically assume I am referring to qualitative aspects of change. It is probably the strong influence of physics and philosophy on these languages which has made it customary to think that the only alternative to changes in quantity are changes in quality. Yet, if one refers to human beings 'changes in quality' is too indeterminate. In fact what changes in the course of a civilizing process is not simply the quality of human beings, but the structure of their personality. It is, to mention two of its aspects, the balance and, indeed, the whole relationship between the unlearned elementary impulses of a person and the learned pattern of control and restraint. As for self-restraint I have already mentioned that no human group, however early in type, lacks either the potential or its actualization. But the pattern of restraint, the whole social mould into which the steering of an individual's feeling and conduct is cast, may be very different at different stages of social development. Representatives of later stages, as I have indicated before, are apt to ignore their own social ancestry and the long process of development leading up to themselves. The forgetting of the past, however, is not to be taken lightly. It has far-reaching consequences. The long-lasting failure to determine the nature of 'time' in a manner which can command consensus is one of the many examples which show it. Let me repeat, self-regulation in terms of 'time' which one encounters almost everywhere in later-stage societies is neither a biological datum, part of human nature, nor a metaphysical datum, part of an imaginary *a priori*, but a social datum, an aspect of the developing social habitus of humans which forms part of every individual person.

33 One has to take some care to distinguish between aspects of human beings which do not change over time because they are – or are connected with – biological universals, and other aspects of humans which have not changed because they are closely connected with social problems that have so far remained unmanageable and unsolved, though there are no reasons to think that they cannot be solved. An example of the first kind, of the unchangeable equipment of human beings, is the tendency to respond to conflict situations with what has been called an 'alarm reaction'. It is a reaction which, in broad outline, is shared by the human species with other species. In the case of an experienced danger, of a conflict with living or non-living things, an inborn automatism puts the organism in a different gear, preparing it, as it were, for fight or flight. This is a reliably investigated[19] inborn reaction-pattern which may well give rise to the notion of an inborn aggressiveness. In fact, this automatic change of gear preparing the organism for fast and energetic action appropriate to a danger situation is far less specific then the concept of aggressiveness suggests.

One has to distinguish clearly from such biological universals, the long-standing custom of human beings to settle inter-tribal or inter-state conflicts by reciprocal killings known as 'wars'. There is no evidence for the suggestion that human beings are unable to solve intergroup conflicts by other means than by wars. In fact, at the end of the twentieth century the only question, admittedly a rather important question, appears to be whether they find non-violent means of solving inter-state conflicts before another great war begins or whether yet another war is necessary to find other means of solving inter-state conflicts. But an element in every pre-war situation, a piece in the puzzle of a recurrent drift towards war, is certainly the social habituation of human beings to the solution of inter-state conflicts by means of threats or tests of violence. For many people today the dynamics of inter-state relations which forces humanity to live continuously under the threat of war is as incomprehensible as the continuous pressure of time to which they are exposed. I have tried to throw some light on the conditions of the social personality structure responsible for the time experience of the later-stage societies of our age. Comparison with the time experience of earlier-stage societies have perhaps helped to bring the differences and thus the characteristics of different stages into better relief. But the development of the human time experience should not be seen in isolation. I should like to provide a fuller understanding of it by inviting comparisons with other aspects of earlier societies and the role played there by the social habitus of people.

34 I choose as an example a report of a French missionary, Joseph-François Lafitau, about some of the customs of American Indians which he published in 1724.[20] One could well imagine that the ancestors of the group of which it was said before that they had no word for time, lived in a manner not unlike that of the Indians described by Lafitau. The absence of a generic concept of time, in this case as in others, does not exclude the existence of concepts representing a lower level of synthesis which members of later societies would probably classify as time related concepts. American Indians of the eighteenth century, like other people at the same stage of development, could almost certainly recognize without difficulty whether an animal track was fresh or old and probably

also how old. They may have used special words for a fresh scent and an old scent. When they went on one of their war-like expeditions, intended to raid an enemy village, it might have taken them four or five days to get there. Although they knew how to live on the land they took with them some provisions commensurate to the length of the journey. Perhaps they had a word for journeys of different length. But the Iroquois and hundreds of other tribal groups thinly populating the vast territories of the Americas did not surround themselves with time-keeping instruments. They were not exposed day and night to an even pressure of 'time' such as that characteristic of people at the later stages of the civilizing process; nor to the corresponding pressure of a pattern of self-restraint covering all their impulses, all their activities from one day to another.

This does not mean, as I have already indicated, that they lacked self-restraint altogether. No sane person above the baby level does. A good deal of confused thinking, at present, obscures the available evidence. It may be worth briefly considering it. Perhaps one should remember that human beings, in contrast to some other species living in groups, do not possess any inborn, automatic restraints against the rising of their own anger and fear in case of conflict and danger. Instead, they are by nature equipped with a capacity for restraining these and other elementary impulses which does not function automatically in connection with an internal metabolism or an external releaser. The capacity for self-restraint remains inert unless it is activated and developed by learning, by specific individual experiences. Early childhood patterning of the biological structures of impulse control, early forms of learning lead to the formation of acquired mechanisms of self control, of socially induced mechanisms which, especially in later societies, are no longer amenable to any conscious control, which become, as it were, second nature. However, the unlearned human capacity for counterbalancing according to learned patterns the organism's more elementary and spontaneous impulses is a unique characteristic of the human species.[21]

It is one of the basic shortcomings of many schools of psychology, and particularly of the school of animal psychology founded by Konrad Lorenz, that this unique characteristic of the biological endowment of human beings is almost completely ignored in their theories. The development of a human person is marked by an intimate fusion of biological and social processes. Unlearned natural growth blends with learned experiential developments in such a way that it is futile to attempt to separate one from the other in what emerges from their fusion. Naturalistic psychology, perhaps in an effort to sustain the status of a natural science, ignores the fact that humans in their development bridge the seemingly unalterable divide between 'nature' and 'society', and thus also between 'nature' and 'culture'. As a result, discussions on such topics as the human self-regulation in terms of time in later-stage societies, or the problem of violence and violence-control in human relationships, drag on without any chance of agreed solutions.

The discussion on violence can serve as an example. There are those who maintain that humans have a genetically determined aggression drive, a concept modelled, as it seems, on that of the sex drive. There are others who maintain that aggressive tendencies in humans are exclusively the result of what is called 'environmental' or 'cultural' influences. Few participants in that discussion appear to consider the possibility that a well-known biological reaction pattern, sometimes known as alarm reaction, which in case of conflict or danger mobilizes the organism for fight or flight may have become in humans more malleable and subject to counteracting controls. The standard discussion on the topic of aggressiveness, as on many related topics, in other words, fails to take into account the interplay between affect and affect-control whose pattern can vary greatly from society to society and, within each society, from individual to individual.

Animal psychologists of great merit in their own field who venture into the field of human psychology usually fail to take into account the biological uniqueness of the human species. One aspect of this uniqueness is the capacity of which I have spoken, a biological disposition for controlling and modifying drives and affects in a great variety of ways as part of a learning process. Hence, one cannot be satisfied with the explanation of a particularly warlike behaviour and sentiment of some human groups, such as the American Indians of the eighteenth century, as the result of a particularly heavy dose of innate aggressiveness. This is one more case of explaining that which has to be explained by means of something that stands equally in need of explanation. If the warring Indian tribes of America were really endowed with a particularly strong aggressive drive, were they not equally endowed with a capacity for controlling that drive? In fact, their social pattern of conduct showed specific forms of self-control of a rather extreme type. As an example their case is rather instructive. It was not a particularly strong biological dose of innate aggressiveness which caused their perpetual war-like conflicts. It was rather an inability to master and to control a – for them – inescapable conflict situation, and to alter the corresponding personality structure which kept the flame of their war-like conflicts alive.

Like most peoples at an earlier stage of development, the American Indians had an extremely elaborate social code of conduct and sentiment which demanded self-restraint on specific occasions. But their patterns of self-restraint and thus their social habitus were in some respects very different from those engendered by the social code of societies at a later stage. Like their mode of timing, the self-restraint demanded by their social code was discontinuous, point-like and bound to specific occasions. It was also in connection with some occasions, far more severe and extreme than any pattern of self-restraint characteristic of the later stages of a civilizing process, and more extreme, correspondingly, were the patterns of unrestraint permitted and, in fact, demanded by their social code. Relatively extreme fluctuations from pleasure to pain and back again are quite often to be found as characteristics of the social code of earlier human groups, and as a characteristic of that of later societies a relatively well-tempered, more moderate type of fluctuations between restrictive self-restraint and controlled unrestraint.

The direction of change in the pattern of self-restraint indicated here is not unconnected with advancing internal pacification, the concomitant of advances from roaming bands to stockaded or hill-top villages, to walled city-states and to increasingly larger, internally more and more firmly pacified, territorial states, to mention only a few stations on the road. At all of them one encounters inter-group killings as a way of settling conflicts and as a traditional social institution deeply influencing the social habitus of people. But in many earlier societies one encounters forms of violence between human groups as a more or less continuous or even as a prevailing condition of their existence and often enough as a way of life.

A rather one-sided picture of pre-state societies may emerge if one confines one's study, as some social anthropologists have done, to some more or less pacified territories. By and large one will probably have to agree with the new view expressed by Pierre Clastres when he wrote:

One cannot think of primitive society... without at the same time thinking of war... In the universe of savages warlike violence appears as the principal means... of preserving each community as an autonomous and unified totality, free and independent of others. War, as the major obstacle which societies without a state oppose to the unifying machine which constitutes of the state, is of the essence of primitive society. That is to say that every primitive society is warlike; hence the universality, which has been ethnographically established, of war in the infinite variety of known primitive societies.[22]

That members of pre-state societies in their pre-colonial condition had to live in almost constant fear of each other's violence has also long been my own conviction. However, reflecting on this matter in developmental terms, one can see more clearly that with the advancing processes of state formation larger and larger populations and areas have become internally pacified and that has certainly made all the difference in the world. But although this development has increased among human beings the revulsion against war, it has not eliminated the reciprocal fears of members of different states that they might be overcome by another more powerful state and, through the threat or the use of violence, might be forced to submit to the will of the rulers of another state. The danger of violence between different human groups had not yet been mastered. War-like violence, in many areas of the world, has become a little more distant and somewhat more intermittent. One need not fear every night that a hostile warrior will appear from the dark trying to rob and kill. But although pacification within states has grown apace, the reciprocal fear and the mutual escalation of threats of violence in relations between states not covered by an effective monopoly of violence still remained largely uncontrollable. Analogous threats and fears were even less controllable at the stage at which human survival units had largely the character of pre-state societies. Perhaps one can understand better the ubiquity of the threat of war among those whom Clastres still calls 'savages' and who were simply human beings like ourselves at an earlier stage of social development, if one remembers that we, too, still live at an early stage in the development of humanity – one should probably think of it as humankind's prehistory where humans are still unable to understand and to control the social dynamics which threatened to drive rulers of different states towards settling their conflicts through the use of force. Lafitau wrote of the Iroquois Indians:

The men who are so idle in their villages, regard their indolence as a sign of glory in order to make everyone understand that they are actually only born for the great things and particularly for war. For there they can put their courage to the hardest tests. War gives them many occasions to show to the greatest advantage all their exalted sentiments.[23]

He goes on to say that they might even leave to the women, who in any case have to look to the food, their other two occupations, hunting and fishing, if they did not provide an exercise that could prepare them for their main purpose in life, for making themselves more terrible to their enemies, more terrible, in fact, than any wild beast. But their enemies evidently had the same aim. The picture which emerges from these and other descriptions is that of a vast field of smaller and larger tribal groups all exposed to each other's raiding parties, all trying to surpass each other in the brutality with which they slaughter their victims and the exquisite ferocity with which they torture their prisoners. In fact, they were tied to each other in the form of one of those double-bind processes[24] which forces each of the groups involved in it continuously to escalate the means of harming others, of striking fear into their hearts as a means of paralyzing their resistance, while at the same time steeling themselves against the terrible pain of the tortures that awaited them, should they fall as prisoners into the hands of an enemy.

Here, in fact, in these societies at a very early level of

social development, one already encounters an inter-group relation, an escalation of frightfulness which, *mutatis mutandis,* is in its structure still closely related to the escalation of frightfulness in the inter-state relations of the age of atomic weapons. There is, in fact, not very much to choose between the torment which people threaten to inflict on each other as a result of radiation poisoning, between the slow and painful death in the aftermath of an atomic battle, and the torment American Indians in the heyday of their independence continuously threatened to inflict upon each other. The difference is that the personality structure of the great majority of people forming the kind of state-societies that threaten each other with the wholesale destruction, with the mass killing and mass poisoning of a nuclear war, is entirely geared to peaceful pursuits and minutely regulated in terms of a finely graded time grid which, as it were, has its counterpart in the individual's conscience, in a pattern of self-restraint that is closely connected with an ever present sense of the flux of time.

The pattern of self-restraint characteristic of an Indian warrior is very different. For one thing, his whole life, at least during the years of his manhood, is entirely dedicated to the pursuit of war. But because that is the case, the double-edged character of the fate and the personality structure of people engaged in the preparation of war, in their case, shows itself particularly clearly. I have already alluded to it. They have to pay a heavy price for the enormous pleasure, promised to them by their social code, they may feel while torturing their prisoners to death in the most horrible manner. They can not only expect to be tortured in the same way by others, if they are unlucky and one of their campaigns miscarries. They also have to prepare themselves for that eventuality from childhood. According to a code of honour apparently shared by all the tribes, which are independent as a result of their common exposure to each other's war-like raids and campaigns, it is shameful for a warrior under torture to show the slightest sign of the pain he feels. To do that dishonours his tribe. Each tribe therefore takes great care to train its male members and perhaps even its women from early days on for this possibility – for the possibility that they may have to suffer an exceedingly painful death and that it is dishonourable to show any signs of one's suffering.

Lafitau reports that five-year-old boys already play a game whereby they press some red-hot material between their hands in order to see whether they can bear it stoically as prescribed by their social code. Most of the ordeals they inflict on each other are ordeals by fire. They put a prisoner on a stake and begin with heated metal pieces burning their lower extremities. They may tear out his nails or wrench off a toe or two, putting them into their pipe or perhaps frying and eating them. If in burning or tearing the flesh from the living body they come across a nerve, they may tear it and twist it to increase the pain of the prisoner as well as their own enjoyment. Not all of course can bear this torture in silence. But some apparently, in the heroic manner which Lafitau praises as reminiscent of the ordeal of Christian martyrs, sing the songs of their people or songs reporting their own heroic deeds. They may incite and provoke their tormentors who swarm around them by reporting how much better they have been able to inflict pain on their victims and describe their own greater art of torturing others. After some hours the fury may have diminished. One may allow the prisoner to lie down and rest for a while. After a while he is taken back and the victorious tribe goes on torturing him, burning and tearing him to pieces limb by limb. After one or two days little may be left and someone may give him the final blow which kills him.

There are probably not many people among those

brought up in the most complex and internally most highly pacified Euro-American state-societies who would be able to kill others in this painfully slow manner and enjoy doing so, or who would be able silently to bear being killed in that manner. In fact, Lafitau himself reports that few French or English people captured by Indians and subjected to tortures of this kind were able to bear their suffering as stoically as most of the Indians. Also, there are probably few people among those living in the complex Euro-American societies who are able to imagine that they themselves, or for that matter any human being, could experience the natural and social levels of the world where they lived without the very keen sense of time, without reference to watches and calendars serving them as auxiliaries of their self-regulation and their self-restraint. But one is not usually aware today that by penetrating to this irremovable furniture of one's whole person, one penetrates to the social layer of one's individual personality. It seems so firmly rooted there that it appears almost as part of one's nature. The same goes for the normal revulsion against putting people to death in the manner of the Red Indians. That too can appear as part of one's own or of human nature generally, while, in fact, it is the result not of nature but of breeding.

The same goes for what we see as the savagery of the savages. It is certainly tempting to classify and, hopefully, to explain the wild pleasure of many American Indian tribes in torturing others in psychiatric terms as a form of sadism or in biologistic terms, for example, as expression of an innate aggressiveness. But these are sham explanations. It is the explanation of an *explanandum* in terms of another *explanandum*. At first sight one may have the impression that the tormenting of prisoners is simply a manifestation of spontaneous impulses. On closer inspection one discovers that it forms part of a highly formalized ritual. The council of elders has in secret decided which prisoners are to be tortured and killed and which are being allowed to live in a bereaved family as a substitute for a member it has lost in battle. Often enough prisoners are left a certain freedom and deliberately deceived about their final destination before they are led to the torture stake. It is the social code of simpler human groups such as these which allows and encourages a higher degree of pleasure in the torturing of others as well as a higher degree of self-restraint on being tortured, on being made to suffer pain. All in all one can say that almost all earlier peoples through their social codes impose on their members partially very severe forms of self-restraint and that, at the same time, their social code, oddly enough, leaves room for and, in fact, encourages what are, compared with ours, rather wild and uninhibited forms of enjoyment. What has to be explained in the first instance is not the individual conduct or experience *per se*, but the social code fashioning individual conduct and sentiment; it is, in other words, the social personality structure out of which forges itself a more or less highly differentiated and distinct individual personality.

35 The time-centredness of members of more complex urbanized societies, too, forms part and parcel of their social code and their social personality structure. To bring this fact into fuller relief was the intention which led me into this excursion about the social code and the personality structure of warriors in societies at an earlier stage of development. Admittedly the subject matter has carried me away, it has led me too far away from the problem of time. The social code in the simpler societies contains few time-signals, and those few are all related to specific occasions; none of them approaches the ubiquity and the high-level synthesis characteristic of the time-signals of members of industrial nation-states. The warrior personality of the North American Indians and of the more

southerly Chaco, as described by Clastres, with its enormous fluctuations between cruel pleasure and cruel suffering and its very high centredness on the present moment, can serve as a contrast-picture making it easier to perceive aspects of one's own personality structure such as the ever-present sense of time, with a degree of detachment in its wider context. The contrast may make it easier to connect the common time experience with the stage of social development characteristic of relatively advanced industrial nation-states in particular with the stage of the civilizing process represented by the restrained pattern of their members.

Also the comparison with the personality structure of members of societies at an earlier stage of development brings into focus a problem that is perhaps somewhat neglected. The American Indians are biologically our equals – members of the species *homo sapiens*. An unintended social process has placed them in a position where many of them, in particular their warriors, find meaning in fulfilment in war and torture and in their own heroic silence when they themselves are being tortured. They have not chosen this mode of life; a long blind social development has forced it on them. The same is true of our own mode of life. That we cannot escape from a sense of time or, for that matter, that we are in danger of destroying each other in an atomic war and, so far, cannot find a convincing way to control the danger, is equally symptomatic of the unplanned force of the development of societies. From a distance we may be able to recognize the blind social process which has driven the American Indian warriors into the bottle-neck of their perpetual wars and their reciprocal tortures. One may even be able to recognize with a little detachment that these people are as helplessly exposed to the vagaries of a blind social process that has driven them into their corner, as human beings were once exposed to the vagaries of natural processes, such as floods or epidemics which they could not control. The blind power of natural processes has in no way been extinguished, but it has been subjected to a considerable measure of control in connection with the growing knowledge of these natural processes and of their explanations. At present, human beings have still little awareness of the fact that they themselves, through the intertwining of their activities, set and keep in motion processes which are hardly less unplanned and unintended than natural processes. The highly developed sense of time, discussed in these pages, is an example. It is an aspect of the development of societies which no one planned or intended to come about. We have to live with it or for that matter with a drift towards war, just as the members of simpler societies of whom I have spoken before have to live with the compulsion of their social pattern. As children we, like they, slip into these patterns. They become our own. We may even find them meaningful. Time is a good example of such a pattern. We have slipped into an ever-present sense of time. It has become part of our own person. As such it becomes self-evident. It seems that one cannot experience the world otherwise.

What I have said may help towards some detachment not only with regard to time or for that matter to war, but also with regard to the blind social processes which produce data such as these. Their nature is at present not very well understood. The capacity to control them is in the early stages. Many aspects connected with this subject matter are highly controversial. The subject of time provides a relatively neutral ground and also a good deal of evidence for the on-going discussion. Perhaps one can also gain a little understanding of the cognitive value of a comparative procedure. With their help one can see one's own social habitus and some of the distinguishing characteristics of the stage of development represented by oneself in fuller relief as a result of a confrontation with representatives of another stage and with their social habitus, while the latter stands out more clearly as a result of its confrontation with a later stage. It is from comparisons and confrontations such as these that, in the course of time, a fuller picture of the development of humanity may emerge and thus of the sequence of its stages. The strong tendency towards a withdrawal of social scientists from past and future to the momentary present, stands as much in the way of comparative inquiries as the perception of the past as 'history'.

I have already said that one's own time-experience can become understandable for oneself only with the help of a reconstruction of the past, only by means of a confrontation with former stages of timing, and these in turn become more understandable, as one may have seen, if they are presented as steps on the developmental ladder. In a small way the classical philosophical literature on time is a good example of the confusion that results from a denial of the social past from the incomprehension of the long process which has led humans to where they are now. At the present stage in the development of humanity some of its representatives have the power to destroy each other, themselves and perhaps humanity altogether. In such a situation it is dangerous not to be completely sincere, not to make an effort to shed disguises, idealistic, materialistic or whatever. They are a luxury one can no longer afford. The danger is that the present civilizing spurt has not reached the stage where individual self-restraint takes precedence over restraint by others.

36 One more foray into the past may help to round off the picture. It shows timing in its social setting at an earlier stage. There is some danger that the study of time may be regarded as a specialism, that one may try to explore the development of timing in isolation. I have already indicated that it can be useful to keep an eye on

the development of violence and violence-control even if one is mainly concerned with the problem of time. It may be still more advantageous for an understanding of the nature and function of timing in earlier societies if one can see it as nearly as possible in its original human setting. However, at the present age not many pre-state societies are left which are so unaffected by the tendency towards state-formation in its contemporary form that timing is practised there in one of its traditional forms. Increasingly, the centre of violence-control as well as of time-control has shifted from the pre-state or village state level to the level of states with a permanently specialized governmental, administrative, judicial and military force. One has to rely to some extent on written sources if one looks for authentic memories of life in a traditional village as it was just before, or just at, the time at which the centre of power moved to the level of states with a permanent, specialized staff.

One informative document of this kind, Chinua Achebe's great novel *Arrow of God*,[25] can serve as an example. A conflict about timing is one of its central themes. It provides a first-hand account of life in an Ibo village of Eastern Nigeria at the period of time at which the old way of life though still largely intact was just beginning to change under the impact of the new colonial rule. Achebe's great story unobtrusively but vividly shows the level of violence in everyday life between different village states as well as within a group of villages united under a god and thus with some characteristics of a federal village state, a kind of republic ruled by the assembly of elders and other men of standing, including the Chief-Priest of Ulu, the highest god of the six villages whose rule is the main bond of the federation. His priest, Ezeulu, is the chief figure of the story.

Ezeulu is the model of a proud and powerful priest. Some generations earlier, the six villages, one of which is that of Ezeulu, were greatly plagued by the recurrent attacks by slave raiders.[26] To ward off their attacks the six villages had formed an alliance. The god Ulu and his priest, an ancestor of the present Ezeulu, had become symbols of the unity of the six villages. And the god had fulfilled the villagers' expectation. The foreign warriors were driven off or stayed away. Ulu had proved a strong and powerful god and the god's power legitimized that of his priest. It was he who in some respects regulated the life of the six villages. In the name of his god he determined for the six villages what we call time. He knew how to look out for the new moon. He announced to the people of the six villages that she had indeed been seen, that the new moon had been greeted by him as a welcome guest and that people could now undertake all those errands which depended on the new moon's coming. Of course Ezeulu usually had in advance a rough idea as to when one might expect the new moon's visit to the six villages, but the task of actually discovering her in the sky was not always easy. Especially in the rainy season the moon sometimes hid herself behind the clouds. One could never be quite sure whether she had come until one had seen her.

At that stage, as one may see, the coming of the new moon was not yet experienced as a natural, a mechanical, causal event. It was, in essence, experienced as an encounter between animate beings, as a meeting of persons. It is true that anyone with reasonably good eyes in their head might be able to discover a new moon in the sky. We hear one of Ezeulu's enemies muttering about it but ordinary people might easily disagree with each other as to whether the new moon had truly come or not. In that case the regulative function of the moon's coming as a time signal setting in motion specific human activities might be in jeopardy. Moreover, seeing the new moon at that stage is spontaneously experienced as implying that one is being seen by her. Not everybody's face might be welcome to the visitor. It might be safer to leave it to the priest to initiate with his drum the welcome ceremony for the new arrival and to ensure the guest's goodwill, to prevent the guest from becoming a 'bad moon' and at the same time, giving people a time signal regulating their activities.

This is Achebe's description of Exeulu's wives and children contemplating a new moon:

The little children in Ezeulu's compound joined the rest in welcoming the moon...
The women too were in the open, talking.
'Moon', said the senior wife, Matefi, 'may your face meeting mine bring good fortune.'
'Where is it?' asked Ugoye, the younger wife.
'I don't see it. Or am I blind?'
'Don't you see beyond the top of the ukwa tree? Not there. Follow my finger.'
'Oho, I see it. Moon, may your face meeting mine bring good fortune. But how is it sitting? I don't like its posture.'
'Why?', asked Matefi.
'I think it sits awkwardly – like an evil moon.'
'No,' said Matefi. 'A bad moon does not leave anyone in doubt. Like the one under which Okuata died. Its legs were up in the air.'
'Does the moon kill people?' asked Obiageli, tugging her mother's cloth.[27]

Like other priests of African village states Ezeulu had several timing functions with the help of which specific social activities were regulated. As a means of co-ordinating the activities of a plurality of individuals timing always presupposes the readiness of that plurality to submit to an integrating authority. That the authority needed in order to establish *when* specific communal and individual activities are to be undertaken has been for a long time vested in priests is due to their closer links with the spirit world of which for a long time all the heavenly bodies, the main instruments of timing, are emissaries or representatives. That he knew the secrets of timing, that he could with authority declare or even refuse to declare when specific social activities should start, was certainly one of the sources of a priest's and thus also of Ezeulu's power. It was he who, after consulting with his god, decided when the pumpkin festival should take place. At this festival all the villagers, husbands and wives, assembled at the big market place leaving a large free space in the middle. The enormous drum which could be heard in all the villages and which called the inhabitants or sometimes only the elders together in any emergency, began to speak with its loud exciting voice. After a while Ulu's high priest appeared accompanied by his attendants and danced in a ritual manner around the open space left for him by the public in the middle of the market place. Large pumpkin leaves fell down on the place where he danced. People threw them down there. His dance cleansed them of misdeeds they had committed. On another occasion, it was his duty and his privilege, after hearing the voice of his god, to announce when the right time for the harvest of people's staple food, the yam roots, had come. This was equivalent to announcing the new year. People whose store of yams might have become nearly exhausted could fill it again and begin to enjoy their newly harvested food. For all these time regulations, the people of the six villages depended on Ulu and his priest. Ezeulu was very conscious of the power which these and all his other tasks conferred on him.

He was a tall and dignified man, proud and sure of himself in the knowledge of the confidence of his god. Yet, although he did not know it, his own power as well as that of his god suffered erosion through the coming of the white man and, to some extent, of his new religion. The white man had begun to enforce his peace. There were no longer any alien warriors to be feared from whom one needed protection by Ulu. Even the endemic local wars between different village states were now prohibited. And when the elders of the six villages, despite the warnings of Ulu and the fierce opposition of his priest, decided to go to war against a neighbouring village state for the sake of a disputed piece of land, the white man, who came from a country of which no one had heard anything before, marched in with his troops, collected all the guns from the villagers, broke to pieces the greater part in the sight of everyone and took away the rest. Obviously change was in the air.

The proud and independent peasants of the six villages, like Ibo people elsewhere, like the Romans of the early Republic, refused to submit to chiefs or, as they were sometimes called, to kings. Some of them felt, and a few even said openly, that the chief-priest of Ulu was too power-hungry, too ambitious, almost as if he wanted to be a priest-king. The leading man of the faction opposing Ezeulu who would have liked to break his power if he could, was a wealthy peasant called Nwaka. The hostility between the two men had risen sharply over the question of war which Nwaka wanted and Ezeulu, in the name of his god, denounced as an unjust war. In the trial staged by the colonial administration in order to determine who was to blame for the war Nwaka and the other witnesses from the six villages had firmly asserted that they had a right to the disputed land and that the other side was to blame for the war. Only Ezeulu, upright and inflexible as ever, had declared that it was an unjust war and the white man had decided accordingly.

Many people in the villages felt that the priest had let them down. The tension between the priest on one side and Nwaka with a growing number of partisans on the other rose; Ezeulu in his righteousness grew increasingly bitter. His god was a jealous god expecting obedience from his people and inclined to punish offenders. Hence when the time for the harvest approached he remained silent. His priest, not hearing the god's voice, was unable to announce to the people that the time for the harvest had come. Thus the people suffered; a time of great stress and distress for the six villages began. The old stock of food dwindled, new stocks did not come in to replace them. The people of the villages, including Ezeulu himself, his wives and children, had not enough to eat. Those who could afford it surreptitiously bought supplies from other villages. The Christian mission announced that anyone who brought new crops to the Christian god to be blessed there was safe from Ulu or any other heathen god and had nothing to fear from them. Ezeulu, seeing hunger, starvation and suffering around him, was in conflict with himself. He sincerely waited for the voice of his god to tell him that the time for the harvest and thus the beginning of the new year had come so that he could announce the good news to the people. But he also felt dimly that it was his own wrath, his rage against the villagers and his need for vengeance which closed his ear to the god's voice. But he could not tame his unrelenting anger, his need to punish the people for what they had done to him, even though it meant also punishing himself. The blow that opened the knot came like lightning out of the blue. Suddenly and unexpectedly his favourite son died. It was the son he probably hoped the god would choose to succeed him when he died. The blow struck him down. At once the villagers knew and he himself knew that his god had turned against him. They understood that Ulu had abandoned his priest.

All his life Ezeulu had felt his god near him. He was always certain of his confidence. He had bared his heart to him more frankly and perhaps more warmly than a son might do when speaking to his father. Not hearing the god's voice, and thus in deference to him, he had refused to let the old year die and to announce the beginning of a new year. And the god had rewarded him with the death of his beloved son, thus showing clearly that he had turned against him and had taken the side of his enemies who would now try to hunt him down. Without the confidence and the protection of his god, deprived of his meaning and the strength to live, Ezeulu had come to the end of his road. And so, Achebe tells us, had his god. For what is a god without his priest?

37 At the phase of development represented here, time, as one may see, is not yet experienced as a steady, impersonal flow symbolized by the relentless coming and going of the calendar years, which come sliding down from an almost indeterminate, only partially mortgaged future, enter the present of the living here and now only to merge at the end into the immeasurable past. The contemporary time experience is closely linked to the experience of nature as a largely automatic nexus of impersonal events of which the growth and decay of galaxies forms as much a part as the growing and ageing of one's own person symbolized by a steadily rising number of life-years.

Achebe's account, in its unobtrusive way, allows us a glimpse of an earlier time experience. In that phase the conceptual and experiential distinction between past, present and future or between living and non-living things had not yet attained the sharp edge that it has today. At present, perhaps only the fact that some conceptual symbols of inanimate objects are regarded as masculine, and others as feminine, survives as a vestige of a former experience of these objects as a kind of person.

Does this mean that members of earlier societies experienced as animate what we experience as inanimate? It would be easy if one could answer this question with a simple yes or no. The standard label of the earlier mode of experience as 'animism' lends its support to the yes. The snag is that the present use of concepts such as 'animate', or consequently of 'animism', is based on an extremely reliable and reality-congruent knowledge of all that is implied if one diagnoses some objects as living and others as non-living. Above all, one is able to refer to, and to experience in many situations, living and non-living things alike with a high degree of detachment. The certainty of one's knowledge that the lion in a cage at the zoo has no magical powers which will allow him to break out of the cage and to kill a spectator or two, makes it possible to look at the animal calmly and without any great involvement. The concept 'animism' rightly suggests that members of earlier societies often experienced as animate what members of later societies know to be inanimate. However, as a conceptual symbol of a different mode of experience the term animism is not entirely adequate. It does not draw attention to the fact that objects now known to be inanimate, but at an earlier stage experienced as animate, were not experienced as such with the same degrees of detachment, which is characteristic of such an experience at a later stage. Objects such as the moon and the sun were experienced as a kind of person because their coming and going was experienced with far greater uncertainty and a high degree of involvement as a potential source of danger or perhaps of well-being. The question of which human beings at an earlier stage wanted to know the answer, humanity's primary question, was not 'What is the moon or the sun' and certainly not: 'Are they mineral, vegetable or animal?' The question was: 'What does this or that event in the sky signify for ourselves? Is it good or is it bad for us?'

The sighting of a new moon, in Achebe's story, undoubtedly had for the women and children of the priest's household the character of a personal encounter. It was in that sense 'animistic'. But it also had for them the character of an omen. They were personally involved in that event. The tradition of a society at that stage of development provided its members with certain ritual formulae which defined it for them as a personal encounter. On seeing the new moon for the first time people used to greet her formally with such words as: 'your face meeting mine'. The example shows very clearly what is meant if one speaks of an innocently self-centred or of an involved mode of experience. The moon's coming, as one may remember, immediately gave rise to comments regarding the possible significance of her appearance for those who saw her. That it had such a significance was taken for granted. One of the children asked whether the moon killed people and was reassured by an adult who pointed out that the moon was not such a bad person.

38 At this earlier stage, in other words, the whole world centred on one's own village and including what we call nature was still experienced as a unitary world of spirits. At the later stages this mode of experience does not necessarily disappear. But it ceases to be in many areas of people's lives the dominant mode of experience. As such, as dominant mode of experience, it becomes institutionally more and more confined to special compartments of social life. The primary question, the symbol of people's deep involvement: 'Is it good or bad for me or for us?' never ceases to affect the experience of human beings but in more and more areas of human experience it has ceased to be the dominant question. Particularly in the experience of that area we now call 'nature', the self-centred question: 'What is the meaning and purpose of all this for me or for us?' has been driven from its throne and has become unequivocally subordinated to more impersonal and detached questions such as: 'What is the connection between these events?' In the pursuit of questions of this more detached type the reality-congruence of human

knowledge has greatly increased in a number of areas as has the certainty of knowledge; and so, therefore, has the capacity of humans to control events. In the development of humanity, as we know it, there is never a zero-point of realistic knowledge; but these are stages where the body of reality-oriented knowledge, compared with the fund of fantasy–knowledge, is very small. Hand in hand with the growth of the former goes not only the expanding control of events by humans but also a growing certainty which humans have with regard to the connection and origin of events.

The odd thing is that there is relatively little understanding of this gain in certainty. One is as yet, by and large, only dimly aware that reality-congruent knowledge, such as that of which the scientific knowledge of nature is the best example, helps to limit people's fears; it allows people to act with greater certainty as to what they are doing and to live, within these areas of reliable knowledge, with greater security. Lacking awareness of the enormous gains in the certainty of living and the corresponding limitations of fear and uncertainty which are the hallmark of later-stage societies, people also lack awareness of the high level of danger and insecurity which is the lot of earlier-stage societies. In terms of a later stage one might attribute the sudden illness and the subsequent death of a son to natural causes. A father may deeply mourn his loss but his need to blame someone for it, perhaps even himself, will be in most cases limited by the dominant tendency of his society to explain even very sudden illnesses in terms of 'natural causes'. As long as that is not the case, as long as 'nature' is largely perceived as a world of spirits, a voluntaristic element, a wholly incalculable condition enters all one's experiences. One can only guess which spirit or, in a given case, which human being is responsible for another person's illness and death; one can surmise his intention. But if public opinion, in such cases, fastens on an explanation, it can be usually understood in terms of human conflicts and power struggles such as that between Ezeulu and his enemies.

39 One of the reasons for the high level of uncertainty that is characteristic of the knowledge and thus of the life of people forming societies of an earlier type, is the relatively low level of synthesis of their conceptual symbols. People living in societies of a later type, as a rule, are not fully aware that they routinely handle a great number of conceptual symbols at a very high level of synthesis of which the conceptual symbol 'time' is one.

At present, it is usual in such cases to speak of high-level abstractions rather that of high-level syntheses. But the meaning of the term abstraction, like that of the related conceptual polarity 'abstract and concrete', is not entirely clear. I have already mentioned that it is a little difficult to see from what 'time' is abstracted. Is it abstracted from the acts of looking at their watches performed by millions of people all over the world? But all these people, if asked, might say they want to know what time it is. Again, is the concept of nature abstracted from great numbers of observations about natural events? Whatever the answer to such questions, conceptual symbols, such as time, nature, cause, substance, certainly represent a synthesis at a very high level.

But perhaps a contrasting picture is needed in order to see more clearly what that means. Once more, examples from Achebe's *Arrow of God* can be of help. Ezeulu, the high priest, had been summoned to the British District Officer, whose administrative office and residence were situated in one of the neighbouring villages. Because the priest was late the District Officer, ill-tempered and on the verge of a severe illness, had cursorily ordered that the fetish-priest, for the time being, should be put in the guard-room which served as prison, so as to teach him better manners. The District Officer, meanwhile, had been taken to a hospital with a temperature; the indigenous population at Government Hill interpreted the captain's illness as an act of vengeance of the imprisoned priest and as a sign of his power. Ezeulu was, therefore, treated very well. Nevertheless he felt uneasy. The sky over his prison looked alien to him. He said to himself: How could it be otherwise? 'Every land [that is, every village state] has its own sky.'[28]

Today children learn in many societies the standard knowledge that what one experiences as sky and what is now often called space is that within which the earth floats around the sun. It is not different from one locality on earth to another. An overall picture of the earth moving through space around the sun represents a synthesis on a comparatively high level, the result of many observations of detail and many antecedent syntheses on a smaller scale. It also represents a symbolic synthesis that is highly reality-congruent. For people brought up with concepts representing such a realistic synthesis on a high level, it is easy to experience the world accordingly. But because it is easy for them they may forget and may not care to know that in terms of humanity the ascent to that level of synthesis has been an immensely difficult process lasting many thousands of years.

For members of later societies it was easier to view the world in terms of a high-level synthesis not because they were more intelligent or in any other sense 'better' but simply because they were later, later in a sequence of societies where knowledge happened to grow, because they could benefit from the consolidated results of the continuous process of knowledge growth. These results were not accessible to the people of earlier societies living in the past; they were, until now, not accessible to those of earlier societies living at present. I have mentioned before the case of the superintendent of an Indian reservation in the United States who, with an air of superiority, reported that the Indians did not even have a word for time. However, it was in no way his merit that he knew how to handle symbols of a high-level synthesis such as time. Nor was it in any sense symptomatic of any intrinsic deficiency that the Indians, although they undoubtedly used some timing devices in accordance with the requirements of their mode of living, had no overall concept of time.

Members of societies who as *beati possedentes*, benefit from a rich knowledge heritage including many conceptual representatives of a high-level synthesis, have for many centuries tried in vain to solve what was for them the enigma of that possession. Already in antiquity men like Augustine wondered about time. Kant more than a thousand years later found many admirers for the hypothesis that time and space were representatives of an intellectual synthesis *a priori* which meant, in dry words, that this form of synthesis was part of human nature or inborn. It was, as one may see, a classical case of forgetting the past, of disregarding the whole knowledge process leading up to one's own stage, one's own level of synthesis.

Hence it is of some help to get to know and to reconstruct for one's own understanding the mode of experience of human beings at a stage of development where they did not yet experience time and space at the same high level of synthesis as oneself and as unitary throughout. It is not too difficult to understand that for people who looked at the whole earthly landscape of which the sky forms part with an eye for many details, even the sky formed part of the setting of their home-village and that thus the sky there might look quite different from the sky in another village. At that stage people's own human group and, in a given case, their own village formed the primary frame of reference for their experience of what one now calls the universe. In all likelihood the coming of the new moon too, was at an earlier stage experienced with the unreflective self-centredness of high involvement as a visit of the moon to one's own group and one's own locality on earth. If invented, Achebe's story in all these respects is invented with great understanding. He tells us that while Ezeulu was abroad in another village, his youngest son at home worried because he knew that his father was used to welcoming the new moon and, for her part, the new moon was used to being welcomed by his father.

> What would happen to the new moon? He knew his father had been expecting it before he went away. Would it follow him to Okperi or would it wait for his return? If it appeared in Okperi with what metal gong would Ezeulu receive it? ... The best solution was for the new moon to wait for his return tomorrow.[29]

At first glance the youngster's worries, too, can appear as the product of a writer's poetic imagination. Perhaps they appear as the odd symptoms of an alien mentality. In fact, they are a vivid and authentic example of an earlier stage in the development of people's experience of 'nature' and of 'time'. It is still an open question at what stage in the knowledge process human beings came to communicate by means of conceptual symbols fully representative of the fact that it was the same moon and, with certain well-established differences, the same sky which they saw everywhere.

As long as one lacks contrast pictures, it is difficult to perceive one's own experience of time or, for that matter, of nature in sharp relief. One does not easily become aware in that case that one is accustomed to a highly integrated and unitary mode of timing. And yet it is almost obvious. Timing units such as minutes, hours or years which are all interlocked can apply to the boiling of an egg, to global air traffic, to the life processes of a human being, to the development of a state society, to the formation and decay of stars and galaxies as well as to the formation and decay of the whole universe as we know it.

It is characteristic of a timing synthesis at a lower level that a priest can refuse to announce the time for the harvest, as his god has not yet made it apparent to him that the time has come. Hence the new year remains in abeyance; The village remains, in a telling phrase of Achebe, 'locked in the old year'. The specific regulatory activity we call timing is in this case still firmly embedded in the conception of the world as a society of spirits. Relations between people and what we call 'nature' have a much more personal character than they have at a later stage. Our knowledge is still too fragmented to say with certainty when and where for the first time people began to distinguish more clearly and with greater certainty between the impersonal order of nature and the social order of human beings. But it is well known that a determinant spurt in that direction occurred in ancient Greece. There the distinction between that which belongs to the realm of the laws and that which belongs to the realm of nature was for a time firmly established among learned men. Also in Greek mythology appeared, perhaps for the first time, a time god. The name of the ancient god Chronos was also one of the terms for time. It can hardly be without significance for the development of the human experience of time that a god should have lent his name to the concept. What can be said with certainty is that timing and the concept of time cannot be considered in isolation. They form part and parcel of the overall image people have of their world and of the conditions under which they live in it.

In a period where the chains of interdependency, especially in the economic and military sectors, become very long, in some cases even world-wide, time evidently can represent a synthesis at a higher level than it can at a stage where village states formed the principle survival units and the highest effective level of integration. These states, at the time of Ezeulu, were no longer absolutely autarchic. Markets were already a firmly established institution. But each family unit was still largely dependent on its own food production for its food supply. The level of self-sufficiency was still very high in Achebe's Ibo village. Few chains of functional interdependencies linked these villages to the outside world. Those which did exist were very short. Matches, an early symbol of trade over longer distances, as far as one can see, were still a luxury and perhaps not even in use. For their physical survival, for their defence against slave-raiders and other foreigners, or against each other, the villagers, too, had to rely entirely on their own resources. The change came when they were subjected to the rule of a colonial power which was also a step towards their integration at a higher level, a step on the ladder of a state-formation process. It was this relatively high degree of military and economic self-sufficiecy which found its expression in a localized time experience; it showed itself in such modes of experience as that of the new moon as a kind of visitor to one's own village or that of a village god who for reasons of his own did not allow – in that village alone – the old year to die and the new year to take its place.

40 The double movement towards larger and larger units of social integration and longer and longer chains of social interdependencies also had close connections with specific cognitive changes, among them the ascent to higher levels of conceptual synthesis. The advantages of cognitive changes in that direction for societies developing in that sense are not difficult to see. Such a development allowed the perception and the symbolic representation of more comprehensive and more complex interdependencies in

society and nature. The cognitive mastery of connections over longer distances in time and space was an indispensable aspect of a lessening fantasy content, of an advancing reality-congruence of knowledge. It allowed a steady expansion of human control over non-human nature and thus a lowering of the danger level in that area, although on occasions it also helped to increase the danger which human beings consistuted for each other.

There were, however, losses as well as gains connected with such a development. Not understanding the long ascent to a higher-level synthesis and its conditions, people often get so used to communicating with each other in terms of one of its manifestations, of what they themselves call 'abstractions' at a very high level, that they lose sight of the symbolic representations of sensory details to which all high-level abstractions are related. It is not difficult to discover circles of learned people who communicate with each other in a language full of ritualized high level abstractions which no one outside that circle can understand. Within the circle these ritualized abstractions are used with a halo of associations which are never explicitly stated and which need not be stated because in in-group usage the implied associations are known. Non-initiated people assuming that the inquiries of the select must have results that have cognitive value and relevance for everybody, are often at a loss when they are exposed to a hail of abstractions of this kind. They are unfamiliar with the implied associations with which they are used by the initiated and look in vain for any connection of these abstractions with symbols of anything tangible, with a conceptual synthesis, representative of observable details. Without recognizable links with the latter, symbols of high-level syntheses are often little more than empty words. Getting lost in a maze of symbols of this type is one of the permanent pitfalls of life in societies with a fund of knowledge rich in symbols of high-level syntheses. There are many others.

One may be able to see the balance of gains and losses more clearly if one looks at one of the characteristic means of communication which is widely used in earlier societies. There it is indispensable: it fulfils in specific situations an important function. But it becomes marginal in later societies, where conceptual symbols at a much higher level of synthesis take its place. I refer to the use of proverbs. To members of later societies they may appear as part of the folklore of their forefathers, as a literary form which belongs more to the past than to the present. In earlier societies, at certain stages of their development, proverbs are irreplaceable instruments of communication. Members of these societies use them as a normal vehicle of conversations and discussions much in the same way in which members of later societies use symbols representing a higher level of synthesis often called abstractions or generalizations.

Achebe's *Arrow of God* offers a wealth of examples. In contrast to a collection of proverbs, moreover, he often shows the use of a proverb in a particular situation. Thus Ezeulu, the high priest of Ulu, rejects the accusations of his enemies in the village because in one particular matter he has taken the side of the British administration against his own village. They wanted war and he and his god saw that the war was unjust. At the meeting of the elders he lashes out against them, he reminds them that they had helped the white man to find the way in this country. They should not come to him, Ezeulu, and complain that the white man does this and does that; and he clinches his argument with a proverb: 'The man who brings ant-infested faggots into his hut should not grumble when lizards begin to pay him a visit.'[30] At a later stage of conceptual development one could of course still say in that situation: 'People who complain about something undesirable, should first mind their own business. But one could also say, before they blame others they should first see whether it is not their own fault. The later version relies on more impersonal generalizations, on symbols of a higher-level synthesis. One can see the gain and the loss. Impersonal generalizations, embodying a higher level of detachment, can be far more precise, far less equivocal; but they also tend to be rather inflexible and emotionally dry. Proverbs as a means of communication are in a way open-ended; they are less precise and more equivocal. They often receive an unambiguous meaning only in and through the situation in which they are used. But in a specific situation they are often strikingly apt; there they may leave no doubt as to the speaker's meaning and are often more suitable to clinch a point than symbols of a higher-level synthesis. They are certainly better able to transmit to others a speaker's strong feelings and to evoke directly in others a strong emotional response.

Thus, at a higher level of synthesis people might use words such as: we know when friendship turns into enmity. At a lower level one could use the proverb: 'When a handshake goes beyond the elbow we know it has turned to another thing', which is obviously the saying of a society very familiar with the elementary strategy of hand-to-hand fighting. If an acquaintance passes in a hurry without more than a nod, an appropriate high-level generalization might be: 'He sure must be after something.' At an earlier level one might say somewhat contemptuously in a corresponding situation: 'A toad does not run in daytime unless something is after it.' The imagery of proverbs may show an earlier level of revulsion, and also of contempt, with regard to the products of digestion. A proverb used several times in *Arrow of God* with a number of variations is this: 'The fly that struts around on a mound of excrement wastes its time; the mound will always be greater than the fly.'[31] In the language of higher 'abstraction' one might speak of a man pretending to be greater then he is. If one looks for imagery one might say: 'Like the critic who thought Shakespeare a bad playwright.' In Achebe's story, Ezeulu uses this proverb with reference to his enemies.

Ezeulu is certainly one of the great figures of literature. He believes in Ulu, his god, as he believes in himself, unconditionally and without any reserve. A friend of his speaks of him as being part spirit, part man. He himself believes that he has gifts beyond those of ordinary human beings. He can see into the future. His god speaks to him and to no one else.

He is tall, dignified and, for his contemporaries, often inscrutable. One can see him performing a solemn ritual dance in front of a shrine before the assembled villagers taking away their misdeeds with the pumpkin-leaves they throw down. He can control his words, but not the anger behind his silence, not his passion when he is offended. Ready to attack, he is dangerous like a crouching leopard when attacked. Also he is stubborn to a degree, totally unrelenting in his anger and unable to give way to a degree beyond any bounds of reason until his god betrays him.

In many ways Ezeulu is the model of an early charismatic priest. He will say: 'Ulu and I' as if he were the god's twin. Moreover, his god is still very much an individual person with inexplicable whims and a personal name, not merely a rather abstract label, with the generic name 'god'. Great and powerful priests who felt they were a god's mouthpiece, felt that their voice was the voice of their god, are known from the past no less than from our own time. Achebe's evocation of such a person is a rare achievement. The setting of the story is an early colonial African village-state; it is there that Ezeulu has his roots. The colonial power plays a crucial but still a marginal part in the story. The Europeans act according to their rules, the Africans according to theirs. Their relationship is marked by an almost total incomprehension on both sides. The English administrators still dress up for dinner regardless of the heat; their tradition bears the stamp of a later stage – that is the root of their power superiority – and, representing a different stage, it produces a different personality structure. Their representatives cannot perceive Ezeulu's dignity, nor his greatness; they see only the primitive fetish-priest probably performing disgusting rites for his savage god.

In a way, thus, Achebe's story brings to life not simply an earlier phase in the development of timing, but also its social setting, without which this phase of timing, is difficult to understand. I have often referred to the intermittent and non-continuous character of earlier timing. At the level of the village-state, there are considerable stretches of people's lives for which no timing is needed. One welcomes the moon when she is new because that is traditionally a break in one's own social life; but then the moon can go on her errand, whatever that may be. One uses market-days as a time-meter, saying something like 'three markets ago', where one of a later society might say three weeks ago, again because markets form easily remembered breaks in the social life of the villages. At a later stage one is hardly aware any longer how difficult it can be to communicate in terms of the time moving between two point-like events rather than in terms of these events themselves – in terms of months rather than in terms of new moons, of weeks rather than of markets. One may not be aware that compared with markets the concept of a week represents the ascent to a higher level of synthesis. The same can be said of concepts such as month or year.

41 Perhaps one should add that there is no lack of examples for such an ascent to a higher-level synthesis in the knowledge development of humanity. One striking example is the ascent from the ancient Mesopotamian and Egyptian to the Greek level of mathematics. What in the

Greek tradition became known as the discovery of Pythagoras, the discovery of the universal relationship between the size of the three squares on the sides of a right-angled triangle, had been well known to learned people in Babylonia and Egypt long before Pythagoras was born. But they perceived the regularity at a lower level of synthesis; they presented it by means of numerous examples which could be used for practical ends, for example, by builders of houses. As far as is known, the ancient Babylonians and Egyptians never ascended to the level of synthesis where the many practice-related examples of this regularity could be symbolically represented as special cases of a general regularity. They did not develop the relevant symbols to a higher level of synthesis where a simple mathematical equation was enough to represent this regularity, a symbolic representation of this regularity later well understood and used, as it seems, by learned people in Greece.

Those endowed with the knowledge of a later stage may be tempted to ask: 'Why were the Babylonians in their handling of mathematical symbols unable to ascend to the same level of synthesis later attained by the Greeks?' That is no doubt a legitimate question. However, one cannot explore it without considering the level of knowledge, and thus also of symbol development, from which the Mesopotamian peoples started their advance towards mathematics. Nor can one explore this question if one does not understand the inherent difficulties of developing communicable symbols from a given lower to a higher level. Without explorations of this kind, the question: 'Why were the Babylonians unable . . . ?' can represent a naively self-centred approach. It can mean that, by asking it, one simply looks from the standpoint of oneself or, more generally, from the later stage to the earlier stage. In that way one obscures the fact that humanity, or any one of its subdivisions with the character of a knowledge-continuum, had to pass the earlier stages before moving into the later.

Instead of allowing one's imagination to proceed from a later to an earlier stage, it is more appropriate and productive to proceed from an earlier stage to a later. Compared with its antecedents, with an earlier stage in the development of symbols, the mature form of Babylonian mathematics, in all likelihood, represented an advance. It has the appearance of backwardness only if a later stage in the development of symbols is used as a yardstick for one's assessment of an earlier stage. By reconstructing the actual order of succession, it becomes easier to recognize that the Greek breakthrough in mathematics and some other scientific fields, the breakthrough to a higher level of synthesis, presupposed the antecedent advances in these fields made in the ancient Near East.

For those who already know, it is not very easy to restore for their own understanding the experience of those who did not yet know. If one belongs oneself to those who already possess a fund of knowledge embodying a higher level of synthesis, one can easily overlook how difficult it was for the generations brought up with the conceptual symbols and the whole idiom of knowledge of an earlier level, to develop and to comprehend the symbols and the idiom of a next higher level of synthesis.

42 An example from our own time may make it easier to gain understanding of these difficulties. One should not be left with the impression that such difficulties are confined to the past. In fact, this inquiry itself is a case in point. It leads beyond the present level of discussion about the problem of time. This level is marked by the polarity between naturalistic–philosophical and historical approaches to that problem. Perhaps one can understand better the aim and function of this inquiry if one perceives it as a transitional step from an earlier to a later level of synthesis, from a 'systematic' static or a short-term to a historical long-term developmental approach to the prob-

lem of time, equally remote from philosophical absolutism and from historical relativism. What has been said before about the innocent naturalism of the solution which philosophers such as Kant suggested for the problem of time, need not be repeated. But the difference between a developmental and a historical approach to the problems of time and of the human past generally deserves at the end a few comments.

The main basis of the historians' claim to scientific scholarship is the reliability with which they extract and present from a variety of sources detailed residues of the past. Compared with the historiography of former ages the presentation of past and present events in one's own society or in others with the rigorous attention to the reliability of one's detailed evidence characteristic of modern historiography has been a great advance. It has induced those who study societies in the form we call 'history' to bring to light a growing body of detailed evidence for each of the periods they distinguish between pre-history and the present. As a result, the picture of these periods has become much less speculative; it has become more realistic.

However, while the historians' attention to detail is subject to an exacting professional control, their task of fitting the mass of fragmented details together in the form of a coherent picture is far less rigorously controlled. The historians' synthesis has the form of a narrative linking the certain details to each other in an imaginative but far less certain way. The latitude for the intrusion of personal beliefs and ideals into the historians' narrative is great. It is professionally accepted as a normal practice of historians to apply criteria used for judging contemporaries to groups and individuals of other ages. Nothing is more common than historians sitting in judgement over people of former times, whilst using as a yardstick values of their own time. They thus give us the impression that no essential differences, no changes in the level of development, exist anywhere between pre-history and the present.

This combination of isolated details for which testable evidence is offered and a synthesis of these details which is to a considerable extent an imaginative and untestable invention, imposes specific limitations on the presentation of the human past in the form of 'history'. As one approaches the present from the distant past, sources of evidence tend to become more and more numerous and so do, therefore, the detailed fragments of the past potentially relevant to the historians' work. That is one of the reasons why the representation of the human past in the form we call 'history' mostly has the character of short-term history. Given the great and growing number of details which are brought to light for many periods of the past thanks to the growing number of historians and their labour, each historian's span of work and attention has to be confined to relatively short periods of history. Only for these can students of history claim to be experts. The present division of the long development of humanity into a number of relatively short periods, reflects the historians' concept of their own professional competence. Their professional code and consequently their professional conscience creates a bias in favour of a study of relatively short periods of the past. Their ideal of scholarly expertise is reflected in the vision of a past cut into manageable periods, manageable according to the historians' canon of work. The historians' history, in a word, is short-term history.

For the same reason the synthesis of their material even in the relatively loose form of a narrative, is also confined, as a rule, to relatively short periods. It is, by and large, a low-level synthesis. One historian may be able to give a comprehensive view of Greek Antiquity, another of a period of Chinese history, a third of the Italian Renaissance, a fourth of the history of modern Nigeria and a fifth of North-American history. No common frame of reference, comprehensive and testable at the same time, binds the various 'histories' to each other. Narrative history, in these and other cases, appears to be based on the tacit assumption that all the various periods of the history of peoples in the East and the West, the South and the North, are periods at one and the same level. There are, so it seems, no differences in the stage of development between different periods of history. Although there is no lack of random references to specific short-term developments in the writings of historians, the historical form of reconstructing the past lacks any unifying frame of reference which would make it possible to distinguish and to compare differences in the level of development of different periods and long-term developmental changes within a period.

In spite of its limitations, the presentation of the human past in the form of history has greatly enriched our knowledge. I have no doubt that the historians' work will continue to do that in the future. It would be a grave misunderstanding of what I wish to say if it were regarded as a denial of the cognitive value of the historians' professional work. The intensive study of details of relatively short periods of the past or, in other words, the presently prevailing form of short-term historiography, has an indispensable contribution to make to the exploration and reconstruction of the overall development of humanity. I agree with historians that their work represents a necessary step in the study of the human past. I disagree with their claim that the exploration and presentation of the human past in the form we call 'history', or more precisely 'narrative history', is a sufficient step. I disagree above all with the implied assumption that the symbolic reconstruction of the human past in the form of narrative history is the only and thus the final form of exploring the human past and of presenting symbolic models of past and present in a coherent and testable manner. As provisional hypotheses suggesting how the carefully researched fragments of the past might have been connected, narratives might be useful. But they would be more useful, I believe, if the historians' professional code required a higher level of detachment, if they would use their narrative exposition of past events less freely as a means of fighting ideological battles of their own day.

43 The transition from a historical to a developmental sociological approach to human societies requires the changeover to a higher level of detachment. That is one of the difficulties of such a step. As understood here, models of long-term developments represent a form of symbolic synthesis which is predominantly fact-oriented. The transition to it also represents an ascent to a higher-level synthesis compared with that of a historical narrative. This is another of its difficulties. Recognition of the human past as a development would not be possible without the antecedent step, without the evidence brought to light by historians. Understandably, some of them do their best to devalue or even to block the ascent to recognition of long-term developments and thus to a synthesis at a higher level. As many attempts in that direction have been unreliable and speculative, the resistance is not without foundation. Also, research into long-term developments often requires a break through the period barriers usually limiting the historians' field of vision as well as their competence.

Research into social processes of long duration in accordance with the traditional canon of historical work, may seem to require the mastery of more and more details – more, in fact, than a person can responsibly claim to master. However, competence at a higher level of synthesis does not necessarily require knowledge of a greater number of facts. The contrary may be the case. Students of Babylonian mathematics had to remember a considerable number of cases, whereas students of Greek mathematics could do with remembering a single, the Pythagorean, formula.

In the same way long-term developments can be often symbolically represented if one determines with the help of systematic comparisons between selected samples from earlier and later stages the overall direction of a long-term process. Some sections of this enquiry have been devoted to that task. It was necessary to determine the universal function of timing; in that way one was in a position to know which aspects of earlier and later societies one had to compare in order to discover the overall direction of the development of timing as an activity, as an institution and as an experience. That, to a large extent, has been done here. Factual evidence was needed, but far less than one would need for an unstructured history of time. Even as a contribution to a developmental perspective on time, this is no more than an introductory contribution. As a means of transition that already required an effort. It would have been nice to do more, to determine with greater precision for instance the sequence of stages in the development of timing that lay between the early forms of timing in pre-state societies and the so far latest stage in industrial nation-states.

Older concepts of social development often had a teleological character. The observable direction of every development was understood as a direction towards a goal; and the goal was often regarded as the most important aspect of social development. In this early phase of its development the concept of social development had magical undertones. It was a vehicle of prophecy. It contained the promise of its necessary fulfilment. The goal of social development was a projection of human wishes for constant progress with an ideal society at the end. An investigation of human timing and the experience of time, and of their development, can make it easier to free the concept of social development from the metaphysical residue from its past with which it is encrusted. One can say quite clearly that there is a development from a discontinuous form of timing to a continuous form running round the clock, as it were. But the theoretical model of the development in this direction is to be understood as no more than an expression of its author's desire for a better world. I have no interest in saying that it is better to have the time-experience of later societies. I do not know if it is better. I have simply attempted to solve a previously unsolved problem.

That is not to say that within the development of humankind there are not strands running in a direction which can be defined as progress. Long-term social processes with a dominant trend in this direction are not uncommon. There is a very apt example within the context of this book – the development of the calendar. I should have liked to deal with it at greater length. It illuminates the difference between a historical short-term perspective and the long-term perspective of developmental sociology. The former, through cutting up the human past into individual periods that seem to have a life of their own, hinders or blocks the perception of continuous processes of long duration which do not stop at the frontiers of periods, even if they can be influenced by them. The predominant conception of the past as history favours the perception of discontinuities; it accustoms people to see the past as a plurality of unconnected periods. So strong is this habituation that studies of long-term developments which breach the period frontiers are still quite rare. There is no lack of examples of such processes. Some observations on the development of the European calendar may be enough to illustrate the point.

44 At present a unified calendar is used almost on a global scale. It has its weaknesses. Some people think it would be an improvement if Easter stopped roaming about

in the calendar, if it were, like Christmas, fixed to the same calendar date and if everybody's birthday were attached throughout life to the same week-day. But perhaps a little irregularity is welcome. As a timing device used as a frame of reference for a great variety of human activities, the present calendar serves its function so quietly and smoothly that one often forgets it could be otherwise. One forgets that for thousands of years the calendars people used ran into trouble again and again; they had to be reformed and improved repeatedly until one of them reached the near perfection the European calendar has attained since the last calendar reform.

On a small scale, the development of that calendar is a good example of the long-term continuity characteristic of the development of human knowledge and connected aspects of human societies. The development of the calendar with its vicissitudes, its progressions and regressions, can serve as a small-scale empirical model of a long-term strand in the development of a succession of peoples, overarching short-term historical periods and surpassing the lifespan of some of these peoples. The continuity is brought to one's notice graphically enough if one remembers that the month of August received its name in honour of a Roman emperor and that the Roman god, still remembered when one speaks of a Janus-head, also gave his name to the month of January sometimes, but not always, considered as the first month and thus looking back to the old and forward to the new year. The term calendar itself goes back to a Latin word *calendare* reminiscent of the fact that, in an early period, the priest in Rome, like the priest in the Nigerian village, determined when a new moon had been seen. *Calendare*, to call out, to announce, thus is a memento of the time when in Rome an official went through the streets announcing to all and sundry that the new moon had been sighted and thus a new month had begun.

Perhaps one does not experience as surprising the continuity in the development of knowledge over thousands of years beyond the time-borders of states and of periods which the present calendar represents. Perhaps one does not see that the changes of knowledge which went into the making of the calendar as it is used today had the character of a development. Yet, it is not difficult to discover that people in former times using an ancestral calendar were often troubled by its imperfections, so that they instituted reforms which during the following centuries gave rise to new difficulties and a new reform until the calendar finally reached a degree of perfection, of appropriateness to its social function which ended most calendar problems of former days. Then, when it ran smoothly and bothered no one, one forgot that it was ever otherwise, forgot the millennial development that finally resulted in the adequacy of the human-made calendar symbols both to their social function and to the relevant course of inanimate nature.

The difficulties which had to be overcome can be stated in general terms simply and briefly. But one cannot easily overlook that even a brief statement, which is all that need be done here, cannot fail to show up the problematic character of a long-cherished view. According to a traditional value scale, 'nature' has long been considered as the acme of orderliness, human society appears by comparison as disorderly, if not chaotic. The recurrent troubles of calendar-makers in former days, however, were due to the fact that the course of nature was not regular enough to meet human requirements. As long as people did not possess sufficient human-made timing devices, they used recurrent natural movements – above all those of sun, moon and stars – in some respect as a yardstick for determining recurrent time intervals in their social life. But the apparent movement of the sun to which humans owed such time-units as day and night and the sun-year could not be correlated easily with the movement of the moon to which they owed such units as the months. Human societies on the other hand, as they became more differentiated and complex, required an increasingly precise and unvarying time-regulation.

If tradition required a rotation of state officials, as was the case in Republican Rome, it could become extremely important, especially in a period of power struggles, that the time limits set for entering and for leaving office were publicly known and known beyond doubt. Elections had to be held, taxes, rents, debts and interests had to be paid on 'time'. When Julius Caesar seized power he found that the Roman calendar had got out of gear. In Republican Rome, as in Ezeulu's village-state, the control of time and its public announcement was one of the functions of the college of priests, headed by the Pontifex Maximus. He and his priestly colleagues were the guardians of the state-calendar. As a regulator of social relations, however, the calendar in the late Roman Republic was not yet immune from the effects of power struggles. Interested groups apparently could induce the priests to prolong or to shorten a year. Thus the difficulties of correlating natural and social events combined with the effect of power struggles to bring the calendar into disarray.

Caesar, being virtually a dictator, ordered a thorough reform of the calendar. In the last resort the regulation of social relations in terms of time was in the past always a matter of priestly authorities or of secular state authorities. It was not uncharacteristic of the development of knowledge in ancient Rome that Caesar sent for an erudite man from Egypt, for the astronomer and mathematician Sosigenes, to advise him in his effort at reforming the Roman calendar. The Egyptians already had at that time a long tradition of star observation and calendar making. This was now grafted onto the Roman tradition. The result of Caesar's calendar reform of 46 BC was a calendar with many familiar features, clearly an earlier stage in the development of the contemporary calendar.

The Egyptians had tried to correlate time units oriented by the movements of the moon with the main time unit oriented by movements of the sun, by establishing a year of twelve months each with thirty days, which left five supplementary days to bring in line these months with the sun year. Caesar took one day from the month February and distributed the six supplementary days he now had at his disposal among the six uneven months from January to November, an arrangement clearly foreshadowing that of the calendar of recent times. Shortly after Caesar's death, the month of his birth was named in his honour – July.

Another major calendar reform, as is well known, took place in the period usually called the Renaissance. The Roman Church, a major social channel of continuous knowledge transmission from Roman antiquity to modern times, was largely instrumental in carrying the Julian calendar with some amendments from epoch to epoch. But after more than a thousand years it no longer functioned well. As times went on, Sosigenes' and Caesar's prescription for correlating moon- and sun-oriented time units revealed its inadequacy. While state organization, at least in some parts of Europe, reached a level of efficiency and of internal pacification approximating to that of the *Pax Romana* in ancient times, while this organization together with advances in urbanization and commercialization increased the social need for a public regulation of 'time', the inadequacies of the Julian calendar became more and more apparent.

One of its shortcomings, by no means the only one, showed itself in the fact that the movable holy days, and above all those of Easter, had gradually slipped their moorings. The Jewish tradition had bound the Passover holy days to the sighting of the new moon after the spring equinox. In AD 325 the Council of Nicaea had bound the Christian Easter to the first full moon after the spring equinox. By then its date had already moved from the original 25 March to the 21 March. In the sixteenth century the discrepancy between the human-made symbols of the official calendar and the course of nature represented by the observable movement of the sun had increased by another ten days. From the thirteenth century onwards, the discrepancy had been noted and from time to time Church assemblies had discussed its meaning and possible remedies. But the Church, always slow to break with tradition, had not undertaken anything in the matter until it became very urgent. Finally, Pope Gregory XIII took advice from a Neapolitan physician and astronomer, Luigi Lilio, and after his death from the erudite Clavius. One of their remedies concerned the institution of leap years. They saw that one could correlate the movements of moon and sun better in accordance with human requirements if a leap year occurred only every 400 years. As 1600 was the last leap year, 2000 will be the next.

45 Meanwhile the calender has ceased to be a matter of public concern. The calender reform of 1582 resulted in a better adjustment of the calendar symbols than Caesar's reform to their task of correlating the visible movements of sun and moon in their capacity as time indicators with each other and with the course of social events on earth. In its latest stage, the development of the calendar, like that of other timing devices, has shown signs of a growing detachment of the symbols used for timing from that which they earlier symbolized from the course of nature, from movements of sun, moon and stars. With the help of human-made lighting devices, the night has become part of the day. One may marginally remember that the time-unit of a month used to be closely linked with the waxing and waning of the moon. Apart from experts, few people now

take note of the fact that our year is related to movements of the sun and our month to movements of the moon. People do not mind very much if they live in areas where Easter, the feast of Christ's resurrection, does not coincide with spring, with a new young green breaking through the dark earth, or where Christmas coincides with the spring rains. Human-made symbol-like devices such as calendars and watches are now capable of regulating human relationships in matters of time better than the complicated movements around the sun of the earth and her moon. In this as in other respects humans live more firmly integrated than ever into their self-created universe of symbols. Step by step, in the course of a millennial development, the once-troubling calendar problem has been more or less solved. And as calendars no longer give very much trouble people dismiss from their memory the past in which they still gave trouble. They are not very much interested in the stages of the development in the course of which their ancestors step by step found a solution to the troubling problem. Yet human beings must fail to understand themselves and the possibility of their open future if they fail to integrate into their fund of knowledge that of the development leading from the past to the present.

46 In a small way the development of the calendar is a helpful model showing some of the general characteristics of a long-term development. It can help to de-mystify the concept of social development. As one can see, one of the salient aspects of a development is an urgent social problem with which humans wrestle, which for hundreds, perhaps for thousands of years remains unsolved and perhaps, for the time being, insoluble. The development, progressive or regressive as the case may be, is, in essence, the process in the course of which human groups i.e. step by step and often without being aware of it get nearer to, or further away from, a solution of a problem. As all

social problems are interconnected, studies of single strands, such as that of the development of the calendar, must always be regarded as provisional. But they can show, very clearly, as one may see here, that and why unplanned changes in a particular direction do occur and that one can study them without imputing to them some unknown metaphysical forces or any prophesies implying that every stage towards a solution of social problems will make people happier.

Social developments come to a relative end when the problem generating them is more or less solved. The development of human timing is in that respect a useful example. It shows that the unplanned solution of some problems may itself open up further unsolved problems without necessarily devaluing the advances achieved before. A well-fitting time-grid round the clock is indispensable for the functioning of societies with a growing level of productivity as well as a growing leisure time not purchased at the price of other humans' back-breaking labour and poverty. But the pressure of time, like that of a civilizing code, in its present form creates problems which have yet to be solved. They are likely to give rise to developments on the next higher level.

That most historians so far fail to take account of directional long-term social processes, is in part due, as it seems to me, to a lack of systematic reflection about the problems confronting human groups in past ages as in the present age. People did not and, as a matter of fact, do not always explicitly say: 'That is the problem we have to solve.' Often enough one can see people wrestling with specific problems without naming them and often enough without possessing conceptual symbols for them. In the early stages of the development of timing, peoples certainly had no word equivalent to our 'time'. They would not have been able to express clearly the timing problem with which they were confronted. And yet one cannot do justice to their actions and experiences without explaining their problems in a manner they might not be able to do.

Systematic comparisons may make it easier to do that. They make it easier to understand that the transition from our old to a new level of synthesis is full of difficulties. The habituation to an earlier level tends to block for a while the ascent to the next. However, the transition to a long-term approach to the development of societies, bringing to light, as it does, the order of succession in the sequence of events, is not without its rewards. One can perhaps see it here. What has been done here, goes some way towards showing that the developmental sequence, the succession of stages, is not imposed, as it were, from outside on an inert and structureless historical material, but that it is found in the evidence itself and extracted from it in the form of a theoretical model, of a testable, symbolic representation of the developmental process, as it has taken place.

One may also see that comparisons between different stages of social development and frequent forays up and down the developmental ladder bring to life the manifestation of different stages in a manner not accessible to a short-term, non-developmental and non-comparative historical representation. Much remains to be done. But perhaps one will remember that the moon – which has almost disappeared as a timing device from the life of urbanized citizens of industrial nation-states, who suffer from the pressure of time without understanding it – was once a messenger which allowed people at more or less regular intervals to institute breaks in their social life. Perhaps the foray to Ezeulu's village will make it easier for them to understand their own time experience and thus themselves.

Notes

Preface

Norbert Elias died on 1 August 1990 and was unable to see the final proofs of this essay.

1. Horatius Flaccus, Quintus; The Works of Horace, transl. by David Watson (London, 1792; New York 1976), pp. 196–7.
2. Within the philosophical tradition one usually speaks of an epistemological theory, or a theory of cognition, within the sociological tradition of the sociology of knowledge. The difference between the traditions gives rise to some lack of clarity concerning the definition of knowledge or cognition. It is perhaps not quite superfluous to point out, therefore, that the individual act of cognition is quite inseparable from the knowledge that a person has acquired from others, and so finally from the stage reached by the social fund of knowledge. Someone who knows nothing is incapable of cognition. I therefore do not distinguish here between philosophical and sociological theories of knowledge.
3. Cf. Norbert Elias, *Involvement and Detachment* (Oxford, 1987).
4. Cf. Norbert Elias, *The Civilizing Process*, vol. 2 (Oxford, 1982), esp. pp. 229ff.

Time: An Essay

1. I deliberately avoid speaking of a 'level of abstraction'. For from what is the concept of time abstracted?
2. Norbert Elias, *What is Sociology?* (London, 1978), p. 68.

3   It may be helpful to explain why I am using, in this context, the term 'continuum of changes'. The reason is, briefly, that in many processes of change the unity of the process is due, not to any substance which remains unchanged throughout the process, but to the continuity with which one change emerges from another in an unbroken sequence. Take the example of a specific society, of the Netherlands in the fifteenth and twentieth centuries; what links them to each other is not so much any core which remains unchanged but the continuity of changes by which the twentieth-century society has emerged from that of the fifteenth century, reinforced by the fact that it is a remembered continuity. Take a human being; Hume once confessed that he could not understand in what sense the grown-up person he was now was 'the same' as the little child he used to be. Again, the answer is that the identity is not so much one of substance, but rather of the continuity of changes leading from one stage to another and, in this case too, it is a remembered continuity. What we call the 'animal kingdom' in the evolutionary sense is a continuum of changes: it is the continuity of a multiplicity of changes linking fish to men. The same can be said of the physical universe. There are many other examples. In all these cases it is the continuity of changes which links a later stage to earlier stages.
4   N. A. A. Azu, *Adangbe History* (Accra, 1929), p. 18.
5   The change from 'winter time' to 'summer time' is an example of the exercise of this monopoly in contemporary state-societies.
6   Today one proceeds in exactly the same manner, by changing the words without changing the substance of the argument. Instead of using such terms as 'the *a priori* conditions of human experience', one uses the term 'logical'. This term has now come to be used in a wide and exceedingly general sense, usually with the implication that it refers to regularities of people's reasoning which are prior to and independent of any experience of any learned human knowledge. This extended use has very little to do with the near-mathematical discipline of formal logic, which is a perfectly respectable and fruitful exploration of specific

types of pure relationships. The over-extended use of the term 'logical' while suggesting a connection with formal logic, is thus made to refer to nothing more significant than the postulate that an argument must be consistent in itself, or to the fact that, in human communications, A must always mean A and nothing else, which is often formulated as a tautological 'law of logic'. The prestigious term 'logical', substituting for slightly more old-fashioned terms such as '*a priori*' or 'rational', is used in a not particularly prestigious sense.

7 Censorinus, *De die natali*, Lugduni Bat., 1642, XVI, 3ff. (transl. by the author).

8 This raises an interesting problem which I may be allowed to mention without, at the moment, pursuing it further. Einstein's theory, which stresses – against Newton's 'absolute time' – the view that time (or timing?) is different for different reference units, refers to a specific type of time. In accordance with the dominant tradition of physics, it has a law-like character, i.e. it is based on a specific type of abstraction: whatever is measured is seen as an infinitely repeatable particular case of a general law. The theory of relatively abstracts from, though it can be applied to, unidirectional processes.

If such processes are included in one's theoretical framework, one is confronted with timing problems of a different type. The problem of the connection between 'earlier' and 'later' stages of a sequence is an example. The question is whether an order of succession, such as that leading from a sun-star to a white giant and a red-dwarf phase, i.e. the timing in terms of earlier and later phases of a continuous sequence, also varies in accordance with the movements of different reference units. Shall one say, for argument's sake, that for an observer elsewhere the order of succession might be reversed and that, in relation to him, what we perceive as a later stage of the main sequence of galactic or stellar evolution might appear as an earlier stage and vice versa? Given the theoretical foundations of the model of the main sequence, this seems most unlikely – as unlikely as the idea that observers elsewhere in the universe might see mammals as the ancestors of reptiles, and these as the ancestors of fish.

Thus, for the time being, one might venture to say that the order of succession in such sequences and, hence, the timing of stages as 'earlier' and 'later', is likely to be the same regardless of the position of an observer – though of course the theoretical picture we have at present of the sequential order might prove to be grossly inadequate at a later stage.

In other words, if one could devise evolutionary clocks telling time in terms of changing positions within the evolution of the universe and/or its subdivisions, of galactic and stellar evolution, one would be as near to an 'absolute time' as men ever believed themselves to be. Yet, as soon as that is said, one can see the futility of the expression 'absolute time'. Once more our substantival mode of thinking deceives us. It endows time with an existence of its own. It is quite enough to say that the order of succession in an evolutionary sequence is the same whoever and wherever the observer may be.

9 The recognition of the 'discontinuous' character of timing in the early stages of its development is due to Martin P. Nilsson, who has dealt extensively with the problem in his *Primitive Time-Reckoning* (Lund, 1920). I was halfway through my work before I got hold of this book. Without the evidence presented there I would hardly be as certain of the ubiquity of this structural characteristic of early timing as I am now.

10 I am indebted for this information to Mrs Elke Möller-Korte. She added that this type of clock is quite indispensable when trains run to a tight schedule.

11 Human-centred or 'social' time had its own divisions. Thus one might distinguish between the 'time of the gods' and the 'time of the state', the priests' 'time' and the 'time' of kings or state servants, between 'moon-time' and 'sun-time', etc. But these divisions were of a different type from that between recurrent and non-recurrent time.

12 Galileo Galilei, *Dialogues Concerning Two New Sciences*, transl. by Henry Crew and Alfonso de Salvio (New York

1954), pp. 178–9.
13 The irony of the matter is that, in the thirteenth century, the teaching of Aristotle had been regarded by the authorities of the church as a dangerous deviation. The intellectual movement within the church inspired by the growing number of classical, Arab and Jewish writings which at that time became available in translation was involved in fierce struggles and was repeatedly reprimanded by the orthodoxy of the day. Siger of Brabant, leader of this early wave of European enlightenment at the University of Paris and an ardent Aristotelian, was several times censured and condemned by church authorities and was murdered under obscure circumstances while under some kind of papal detention. Thomas Aquinas, the man of the middle way, did more than anyone else for the reconciliation of the traditional and the new doctrines and, thus, for the absorption of what was known of Aristotle's work into the canon of writings approved by the church as authoritative. He, too, was throughout his life involved in bitter and exhausting intellectual struggles, though in the long run his own and his pupils' views prevailed.

It may be interesting to consider the pattern and timescale of such a long-term process. In its course an innovatory and partly oppressed outsider doctrine became accepted as part of the established teaching; it transformed itself into a traditional and orthodox view and became in turn a weapon of oppression of innovatory outsider views at a new level. In this case the full change from oppressed outsider doctrine to oppressive orthodoxy threatened in turn by a rising outsider doctrine, took about three centuries and a half.
14 It may be worth noting that Tycho Brahe had apparently sometimes used time-meters in order to give greater precision to astronomical observations.
15 G. H. Hardy, *A Mathematician's Apology* (Cambridge, 1948); the quotations which follow, ibid. pp. 21ff.
16 Edward T. Hall, *The Silent Language* (New York, 1959), p. 21 (reprinted Garden City, N Y, 1973).
17 Ibid., p. 25.
18 Elias, *What is Sociology?* pp. 100–3.

19 Walter B. Cannon, *The Wisdom of the Body* [1932], (New York, 1963).
20 Joseph-François Lafitau, *Moeurs des sauvages américains, comparées aux moeurs des premiers temps*, 2 vols (Paris, 1724); selection edited by E. H. Lemay, 2 vols (Paris, 1983). The extract quoted appeared with the title 'De la guerre chez les Indiens' in *Le Débat*, Oct. 1980, no. 5, pp. 60–112.

Lafitau is very careful to suggest to his public that in spite of the horrible things his American savages do to each other they never really overstep too much the rules of French *bien séance*, particularly with regard to nudity. One of the most interesting aspects of his book is the total inability of the illustrator to convey visually anything of the strong emotions one cannot help experiencing if one reads some of Lafitau's descriptions, especially about the torture scenes, although Lafitau himself, no doubt, tempers his description which is evidently selective. He is torn between the desire to give his readers a truthful and vivid picture of what he has seen or heard during the years he lived among the Indians, and the wish both to spare his reader's feelings as far as it is possible without being untruthful and not to let his charges appear in too unfavourable a light to his French public. The illustrations show in an almost touching manner the complete inability of an artist brought up in the more civilized tradition of a court-dominated iconography, of a Baroque artist, to break through the hard shell of his artistic convention. Even at the torture stake an Indian warrior shows his courage by means of elegant gestures from the repertoire of a French nobleman in a Parisian salon.

What we call the artistic style of a period – Romanesque, Renaissance or Baroque – is merely an aspect of the social canon of a human group at a particular stage of its development; like other aspects it becomes deeply ingrained in the personal make-up of individual people. It helps to fashion, but it also limits, the personal taste and the vision of individuals in varying degrees; for the elasticity of social canons, their scope for individual variations, can differ greatly. In simpler societies codes of conduct and sentiment tend to be extremely rigid. They were not quite as rigid in Medieval and modern state-societies though still, as the example of Lafitau's illustrator shows, still firmly binding an individual's taste and vision; and have only more recently become flexible enough to allow a very high degree of individualization, so high in fact that people are apt to forget the common code of sentiment and conduct without which communication would be difficult, perhaps impossible. In Lafitau's time, public tortures of people who were or were believed to be criminals were still quite frequent in France. But the artistic canon largely determined by the taste of French court society was highly selective. It left very little room for the unpolished representation of extreme human situations and the powerful emotions engendered by them. Limitations of this kind have also to be considered in judging the accuracy of Lafitau's report. He was certainly not insensitive to the sensibilities of his public. He tried to endow his American savages with a residue of French decorum. He detected – or believed he had detected – many similarities between peoples mentioned by writers of classical Antiquity and the American Indians which he described in a double-edged manner: one cannot be quite sure whether he tried to elevate his savages or to cut down to size the old pagans of Antiquity. Perhaps with some justification, he emphasized the heroism of these poor victims of savage tortures who had been taught to bear them without flinching, without any sign of the pain they suffered. But though one cannot help being conscious of Lafitau's limitations, he is still a magnificent eye-witness of a state in the development of human societies instances of which are now fast disappearing but which one has to know as clearly and vividly as possible if one wants to know oneself.

21 Some other domesticated species show traces of this capacity. One has sometimes spoken of the self-domestication of the human species. The capacity for learning restraint, as distinct from innate restraint, may well be one of its characteristics. There is, at any rate, no other species with a

comparable biological capacity for socially acquired and developed forms of self-restraint and with the corresponding plasticity of drives and affects.
22 Pierre Clastres, 'Malheur du guerrier sauvage', in Clastres, *Recherches d'anthropologie politique* (Paris, 1980), pp. 209ff.
23 Lafitau, *Moeurs des sauvages*, p. 62.
24 On this concept and its application to relationships between tribal groups and states cf. Elias, *Involvement and Detachment*, esp. pp. 74ff.
25 Chinua Achebe, *Arrow of God* (London, 1964).
26 There is good evidence of the fact that in the course of European slaving activities which on the southern coast of Nigeria lasted from about the seventeenth to the nineteenth century, a number of Ibo villages were completely destroyed and some of their inhabitants transported to the New World. One of the slave raiding groups formed by Ibo themselves and known as Ara Chuku is mentioned under a slightly disguised name by Achebe in his novel.
27 Achebe, *Arrow of God*, pp. 2f.
28 Ibid., p. 196
29 Ibid., p. 205.
30 Ibid., p. 163.
31 Ibid., p. 282; also p. 161.

Index

abstractions 181
  changes from particularizing to general 40–1
  ritualized 180
Achebe, Chinua
  Arrow of God 164–70, 171–2, 174–5, 177, 178, 179, 181–4, 195, 200
action, theories of 18
active timing 50, 51, 53–4, 91
activities
  time as means of co-ordinating 121–2
  timing of 49–52, 132–3
Aeschylus 130, 131
age of individuals
  and calendar time 28–9
  in simpler societies 6, 7–8
age time-scale 69–71
aggression
  in human groups 152–3
  in tribal societies 154–60
agriculture, timing of activities 49–52, 58
alarm reaction 149, 152
Alberti, Leon 112
Amazon Indian tribes 135
American Indians
  concept of time 137–8, 139, 150–1, 175–6
  patterns of
    self-restraint 153–4, 157–60
  social code 160–2
  and war 156, 161
animism, concept of 170–2
Archimedes 130, 131
Aristophanes 54
Aristotle 106, 111, 112
Arrow of God see Achebe, Chinua
Assyria 53
Athens 53–4, 102–3
attitudes to time
  in simpler societies 6–8, 24–7, 136–41
  see also concept of time
Auden, W. H. 136
Augustine, Saint 176

Babylonian era time-scale 57
Babylonian mathematics 185–6, 190
Bacon, Francis 43
Bergson, L. 85
biological time 66, 97

Caesar, Julius 54, 55, 195–6
calendar, the
  development of 192–9

calendar, the (*cont'd*)
  Gregorian 55
  Julian 54, 55, 195–6
  and the knowledge of time 6, 7–8
  reforms 54–5, 194, 195–8
  social function of 55–6
  as time-meter 99
  as time symbol 23
calendar time 16–17
  age of individuals and societies determined by 28–9
Censorinus 77–8
changes
  in the concept of time 39–41
  conceptualization of processes of 67–8
  in space and time 100
  and time relationships 46–8
  timing as sequence of continuous 72–4
Charles IX, King of France 55
childhood, learning self-regulation/restraint 11, 151
Chronos (god) 178
civilizing process
  and the development of the social habitus 32–3
  and self-regulation 24–5, 147, 148
  and self-restraint 11, 33–4
Clastres, Pierre 154–5, 161
clepsydra 114
clocks
  distinguishing characteristics 118, 120
  independent existence of 118
  and the measurement of time 1–2, 3–4
  and social development 199
  and social time 116
  and time 12, 13–15, 16

  and time relationships 46
communal functions of time 3–4
communal life
  and the socialization of the individual 17–18, 19
  and the timing of activities 49–52
complex societies (developed, industrial societies)
  attitudes to death and torture 159
  attitudes to time 159
  learning of time 139–40
  need for timing 122
  patterns of self-restraint 27–8
  sense of time 135–6
  social code 160, 161
compulsion of time 21–2, 22–3
Comte, A. 127
concept of time
  American Indians 137–8, 139, 150–1, 175–6
  changes in 39–41
  earlier and later 74–5, 79
  evolution of 132
  in Kant 63
  past, present and future 76–80, 81, 144
  and philosophers 122–7
  in simpler societies 199–200
  structural and experiential 80, 81, 83
  see also attitudes to time
conceptual symbols, in simpler societies 174–9
concrete timing 92

Darwin, Charles 93
death, fear of 129
Descartes, René 4, 28, 38, 39, 61–2, 63–4, 123

detachment, and
    involvement   31–2, 33
developed societies *see* complex
    societies

earlier, concept of   74–5, 79, 80
Easter, and calendar
    reform   196–7
Egyptians, and calendar
    reform   195, 196
Einstein, A.   39–40, 44, 57, 66,
    78, 85, 99
environmentalism   9
era time-scales   56–8
ethnic differences   137
experiential time   97
external world, divided from
    inner world   124–6
Ezeulu (priest in *Arrow of
    God*)   164–70, 174–5, 177,
    179, 181–4, 195, 200

France, calendar reforms   55
future, concept of   76–80, 81,
    144

Galileo   3, 104, 116, 127, 128
God-centred frame of reference,
    in the Middle Ages   105–7
Greek mathematics   184–6,
    190–1
Greek mythology   178
Gregorian calendar   55, 197
Gregory XIII, Pope   55, 197

Hall, Edward T.   137–8
Hardy, G. H.   130, 131
Hegel, G.   127
Heidegger, M.   85
history
    and the limitations of
        synthesis   186–90
    and nature   74
    and social development   199
Horace   22
hour, concept of   76, 77, 79
hourglass, measurement of time
    by   102–3
human consciousness, time as
    universal form of   123–6
human groups
    capacity for intergenerational
        learning   37–8, 39, 61
    development of, and
        time-experiences   139–41
    idea of conceptless   58–61,
        64–71
    regulation of communal life by
        time   3–4
    and time relationships   46–8
    violence and aggression
        in   152–3

identity
    and the conceptless human
        group   64, 67–8
    individual   69–71
independent existence of
    time   120–1
Indians *see* American Indians;
    Pueblo Indians
individuals
    autonomy of   17
    character formation   27–8
    and the concept of
        time   12–13
    experiences of time   10–12
    identity   69–71
    knowledge of own age   6,
        7–8, 28–9
    lack of identity, in simpler
        societies   7–8
    nature and society   15–17

individuals (*cont'd*)
  personal compulsions
    of 19–20
  self-regulation 11, 13, 23,
    24–5, 145, 148
  social development 141–4
  socialization 17–19
  and society 20–2
  and time as standard
    continuum 47–8
industrial societies *see* complex
  societies
initiation rites 146–7
inner world, divided from
  external world 124–6
intellectual traditions 93–4
involvement, and
  detachment 31–2, 33

Julian calender 54, 55, 195–6

Kant, Immanuel, 4, 38, 39, 63,
  130, 176, 187
knowledge
  development of 3, 32
  external and internal worlds
    of 125
  reality-congruent 172–3
  sociological theory of 30
  of time 5–6

Lafitau, Joseph-François 150,
  156, 158, 159
language
  and the concept of
    time 42–4
  human communication
    through 17
  and social development 142
  and the socialization of the
    individual 18–19, 20
later, concept of 74–5, 79, 80

law of falling bodies 110, 111,
  112–14, 127
laws, scientific, status
  of 128–30
Lilio, Luigi 197
Lorenz, Konrad 152

Masaccio 112
masks, and clocks 118
Mathematician's Apology, A
  (Hardy) 130
mathematics
  Babylonian 185–6, 190
  Greek 184–6, 190–1
  and pure relation-symbols
    133–4
  status of 130–2
  unchanging properties of 127
mathematization of time 107
measuring time 1–4, 8, 9–10,
  12, 14, 46
  in Athens, by the
    hourglass 102–3
  Galileo's experiments 107–15
  by time-meters 103–4
medieval society (Middle Ages)
  God-centred frame of
    reference 105–7
  self-regulation in 147
Minkowski, Hermann 99
month, concept of the 76, 77,
  79, 197–8
moon, the
  and the calendar 195
  concept of 66–7
  and the month 197–8
  perception of, in simpler
    societies 90, 91, 92, 200
  sighting of the new
    moon 165–6, 171–2,
    177, 184

and the timing of
    activities 52
Morgenstern 18

narrative history 187, 189–90
natural sciences
  development of 82–3
  and individual psychology
    142
  and the measurement of
    time 3
  and nature 8–9
  and social sciences 84–9
  see also physical sciences
nature
  and the calendar 194–5
  concept of 8–9, 172–4
  and culture 152
  distancing of men from
    115–16
  and the experience of time
    170
  and history 74
  individuals and society 15–17
  measuring time through
    process of 2–3
  and the medieval concept of
    timing 105–7
  and modern symbols 197–8
  as a realm of spirits 31
  and society 41–2, 85–9,
    96–7, 152
  and theories of time 4, 5
  and time-concepts 79–80
nature-centred manner of
    timing 115
'new realism', and Galileo's
    experiments 112
Newton, Sir Isaac 4, 44, 123,
    129
Newtonian time-concept 39–40

Nicaea, Council of 197
nominalism 125
nuclear war 157

object, and subject 124–5
objectivist theory of time 4, 5
objects of knowledge, time as
    property of 126
ontological status of time 12
orientation
  social symbols as means
    of 31
  time as a means of 3, 13,
    21, 34–5, 38

passive timing 49, 50, 51
past, concept of 76–80, 81,
    144
philosophers
  concept of time 122–7
  and theories of time 117
philosophy
  Cartesian 61–2, 63–4
  theories of time 4–5
physical object, time as 123–6
physical sciences
  and the measurement of
    time 3
  and physical and social
    time 116–17
  see also natural sciences
physical time 114–15
  and social time 8, 44–5, 88,
    96, 97, 104–5, 115–18
physicists, and theories of
    time 117
physics 101
  and the concept of time 45
  and the measurement of
    time 3
  process-models 129
  and time-meters 95

physics (*cont'd*)
  unchanging properties of 128
Pontifex Maximus 54, 195
positional relations, in time and space 99–100
positivism 125
present, concept of 76–80, 81, 144
priests, and the timing of activities 50, 51–2, 53–4, 58, 166–7, 178, 193
primitive societies *see* simpler societies
progress, and social development 92–3
proverbs, use of in simpler societies 181–3
psychology 141–4
  and self-regulation 152
Ptolemy 105
Pueblo Indians, different time-experiences of 137–8
pulse-beat, as time-meter 109–10, 114
Pythagoras, discovery of 185

racial differences 136–7
reason
  in Descartes 61–2
  individual self-restraint as 146
Roman calendar reform 195–6
Russell, Bertrand 130

scientific work, unchanging properties of 127–8
scientists, acts of distancing by 124
self-regulation
  learning through time 11, 21, 23, 24–5
  and social development 144–6
self restraint
  and the civilizing process 33–4
  in developed societies 27–8
  human capacity for 151–2
  in simpler societies 24–5, 146–7, 157–60
  and social development 146–9
simpler societies (tribal societies) 150–84
  attitudes to time 6–8, 24–7, 136–41
  inititation rites 146–7
  lack of individual identity 7–8
  need for timing 122
  self-restraint 24–5, 146–7, 157–60
  social code 160–3
  use of proverbs 181–3
sky, concept of 175, 176, 177
social code in simpler societies 153–4, 160–3
social development 198–200
  and the ideal of progress 92–3
  and self-regulation/restraint 144–9
  teleological character of 191–2
  unplanned force of 161–3
social functions of time 3–4
social psychology 141–4
social sciences and natural sciences 84–9
social symbols 34
  of human groups 20–1
  theory of 132

social time and physical time 8, 44–5, 88, 96, 97, 104–5, 115–18
socialization of individuals 17–19
societies
  age of, and calendar time 28–9
  and individuals 15–17, 20–2
  individuals and nature 15–17
  and nature 41–2, 85–9, 96–7
  see also complex societies; simpler societies
sociological theory of knowledge and perception 30
sociology
  of knowledge 132
  of time 45, 53, 105
solar year 29, 73
Sosigenes 195, 196
space
  concept of 175
  and time 98–100
space-meters 95, 99, 100
spirits, world of
  and nature 31
  in simpler societies 25–6, 178
standard continuum, time as 46–8
state, development of and active timing 53–4
Stonehenge 89–91
subject, and object 124–5
subjective experiences of time 11–12
subjectivist theory of time 4–5
subjects of knowledge, time as property of 126
substantival form of the concept of time 42–3, 46, 73

summer solstice 90
sun, the
  and the calendar 194–5
  and the fixing of the year 55, 56
  and the measurement of time 10, 14
  as timing device 90–1
  and the timing of activities 52
symbols
  conceptual, in simpler societies 174–9
  distinguishing from reality 23–4
  human communication through 17
  pure relation-symbols 133–5
  in simpler societies 26
  of time 14–15, 29–30
  see also social symbols
synthesis
  ascent to higher levels of 184–6
  of conceptual symbols 174–9
  of events 96
  and history 186–90

time-meters 40, 41, 44, 95, 99
  distinguishing characteristics 118–20
  in Galileo's experiments 109–10
  market-days as 184
  social and physical 57
time-regulators 41
time-scales, non-recurrent 56–8
timing devices
  man-made 84
  movements of 118–21

torture, and Indian warriors 157–60, 161
tribal societies *see* simpler societies

Uccello, Paolo 112
universe
    concept of, and simpler societies 176–7
    determining age of 29
    five-dimensional 35–6, 59, 81, 132

verbal form of the concept of time 43, 44, 46, 73

violence *see* aggression

wars 149–50, 156, 157, 161
watches 73, 99
Weber, Max 18
'when'-aspects 71
'wolf', concept of 65–6

years
    concept of 76, 77, 79
    establishing dating 54–6
    regulation of time through 22
    social sequence of 79
    solar 29, 73

63196

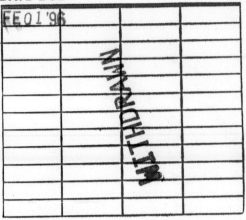

```
BD                63196
638
.E42    Elias, Norbert
1992       Time.
```

HIEBERT LIBRARY
Fresno Pacific College - M.B. Seminary
Fresno, CA 93702